Iron Metabolism - A Double-Edged Sword

*Edited by Marwa Zakaria
and Tamer Hassan*

Published in London, United Kingdom

Iron Metabolism - A Double-Edged Sword
http://dx.doi.org/10.5772/intechopen.95150
Edited by Marwa Zakaria and Tamer Hassan

Contributors
Safa A. Faraj, Naeem M. Al-Abedy, Roberto Miniero, Valentina Talarico, Marco Bertini, Giuseppe Antonio Mazza, Laura Giancotti, Santina Marrazzo, Asuman Akkaya Fırat, Yulia Nadar Indrasari, Siti Nurul Hapsari, Muhamad Robiul Fuadi, Kouser Fathima Firdose, Noor Firdose, Eeka Prabhakar, Nitai Charan Giri, Chinmayee Dahihandekar, Sweta Kale Pisulkar

Notice
Statements and opinions expressed in the chapters are these of the individual contributors and not necessarily those of the editors or publisher. No responsibility is accepted for the accuracy of information contained in the published chapters. The publisher assumes no responsibility for any damage or injury to persons or property arising out of the use of any materials, instructions, methods or ideas contained in the book.

First published in London, United Kingdom, 2022 by IntechOpen
IntechOpen is the global imprint of INTECHOPEN LIMITED, registered in England and Wales, registration number: 11086078, 5 Princes Gate Court, London, SW7 2QJ, United Kingdom
Printed in Croatia

British Library Cataloguing-in-Publication Data
A catalogue record for this book is available from the British Library

Additional hard and PDF copies can be obtained from orders@intechopen.com

Iron Metabolism - A Double-Edged Sword
Edited by Marwa Zakaria and Tamer Hassan
p. cm.

This title is part of the Biochemistry Book Series, Volume 37
Topic: Metabolism
Series Editor: Miroslav Blumenberg
Topic Editor: Yannis Karamanos

Print ISBN 978-1-83962-997-6
Online ISBN 978-1-83962-998-3
eBook (PDF) ISBN 978-1-83962-999-0
ISSN 2632-0983

IntechOpen Book Series
Biochemistry
Volume 37

Aims and Scope of the Series

Biochemistry, the study of chemical transformations occurring within living organisms, impacts all of the life sciences, from molecular crystallography and genetics, to ecology, medicine and population biology. Biochemistry studies macromolecules - proteins, nucleic acids, carbohydrates and lipids –their building blocks, structures, functions and interactions. Much of biochemistry is devoted to enzymes, proteins that catalyze chemical reactions, enzyme structures, mechanisms of action and their roles within cells. Biochemistry also studies small signaling molecules, coenzymes, inhibitors, vitamins and hormones, which play roles in the life process. Biochemical experimentation, besides coopting the methods of classical chemistry, e.g., chromatography, adopted new techniques, e.g., X-ray diffraction, electron microscopy, NMR, radioisotopes, and developed sophisticated microbial genetic tools, e.g., auxotroph mutants and their revertants, fermentation, etc. More recently, biochemistry embraced the 'big data' omics systems. Initial biochemical studies have been exclusively analytic: dissecting, purifying and examining individual components of a biological system; in exemplary words of Efraim Racker, (1913 –1991) "Don't waste clean thinking on dirty enzymes." Today, however, biochemistry is becoming more agglomerative and comprehensive, setting out to integrate and describe fully a particular biological system. The 'big data' metabolomics can define the complement of small molecules, e.g., in a soil or biofilm sample; proteomics can distinguish all the proteins comprising e.g., serum; metagenomics can identify all the genes in a complex environment e.g., the bovine rumen.

This Biochemistry Series will address both the current research on biomolecules, and the emerging trends with great promise.

Meet the Series Editor

Miroslav Blumenberg, Ph.D., was born in Subotica and received his BSc in Belgrade, Yugoslavia. He completed his Ph.D. at MIT in Organic Chemistry; he followed up his Ph.D. with two postdoctoral study periods at Stanford University. Since 1983, he has been a faculty member of the RO Perelman Department of Dermatology, NYU School of Medicine, where he is codirector of a training grant in cutaneous biology. Dr. Blumenberg's research is focused on the epidermis, expression of keratin genes, transcription profiling, keratinocyte differentiation, inflammatory diseases and cancers, and most recently the effects of the microbiome on the skin. He has published more than 100 peer-reviewed research articles and graduated numerous Ph.D. and postdoctoral students.

Meet the Volume Editors

Prof. Marwa Zakaria is an Associate Professor of Pediatrics and Pediatric Hematology and Oncology, Pediatric Department, Zagazig University, Egypt. She is an active member of the International Society of Pediatric Oncology (SIOP), European Hematology Association (EHA), and Egyptian Society of Pediatric Hematology and Oncology (ESPHO). She has participated in several professional trainings and workshops, including ICH GCP online training, EHA Master Class and Bite-size Master Class, and training from the Society of Neuro-Oncology (SNO). She completed a postgraduate training program in Pediatric Nutrition at the School of Medicine, Boston University, USA, in 2017. She completed several international preceptorships, including a thalassemia preceptorship and a hemophilia preceptorship. Dr. Zakaria is the recipient of a 2018 award from SIOP, and scholarships from EHA-HOPE in 2017 and 2018. She has participated in many international and national pediatric and hematology conferences, where she has also been a guest speaker. She has more than forty international research publications in pediatrics and pediatric hematology and oncology to her credit. She has edited three books and five book chapters. She is also a reviewer for several journals, including *Medicine, Frontiers in Pediatrics, Molecular Medicine Reports, International Journal of Infectious Diseases*, and others. Dr. Zakaria served as co-investigator for four hematology clinical trials and sub-investigator for five others.

Tamer Hassen is a Professor of Pediatrics, Faculty of Medicine, Zagazig University, Egypt. He is an active member of the European Hematology Association (EHA), International Society of Pediatric Oncology (SIOP), and Egyptian Society of Pediatric Hematology and Oncology (ESPHO), and has attended numerous national and international pediatric and hematology conferences held by these organizations and others. He has been a guest speaker at numerous pediatric oncology and hematology meetings and has published more than fifty international research publications in pediatrics and pediatric hematology and oncology. Dr. Hassan has edited two books and authored four book chapters. He has participated in many professional trainings and workshops. He received international scholarships from EHA-HOPE Cairo in 2017 and 2018, and an award from SIOP in 2016. He has completed several international preceptorships, including a hemophilia preceptorship at Saint Luc Hospital, Brussels, Belgium, and an immune-thrombocytopenia (ITP) preceptorship at Dmitry Rogachev National Research Center of Pediatric Hematology, Oncology and Immunology, Moscow, Russia. Dr. Hassan is an editor and reviewer for many journals, including *Hemophilia, Medicine, Oncology Letters, Child Neurology*, and more. He was a primary investigator in four international clinical trials and a sub-investigator for ten others.

Contents

Preface

Iron is a vital trace element for humans, as it plays a crucial role in oxygen transport, oxidative metabolism, cellular proliferation, and many catalytic reactions. To be beneficial, the amount of iron in the human body needs to be maintained within the ideal range. Iron metabolism is one of the most complex processes involving many organs and tissues, the interaction of which is critical for iron homeostasis. The bone marrow is the prime iron consumer in the body, being the site for erythropoiesis, while the reticuloendothelial system is responsible for iron recycling through erythrocyte phagocytosis. Among the numerous proteins involved in iron metabolism, hepcidin is a liver-derived peptide hormone, which is the master regulator of iron metabolism. This hormone acts in many target tissues and regulates systemic iron levels through a negative feedback mechanism. Hepcidin synthesis is controlled by several factors such as iron levels, anemia, infection, inflammation, and erythropoietic activity. In addition to systemic control, iron balance mechanisms also exist at the cellular level and include the interaction between iron-regulatory proteins and iron-responsive elements. Genetic and acquired diseases of the tissues involved in iron metabolism cause a dysregulation of the iron cycle. Consequently, iron deficiency or excess can result, both of which have detrimental effects on the organism.

This book contains eight chapters divided into three sections on iron, iron metabolism and iron deficiency and new lines of therapy.

Nowadays, research is increasingly dedicated to the field of iron metabolism, but countless questions remain unanswered. Increased knowledge of the physiology of iron homeostasis facilitates understanding the pathology of iron disorders, such as iron deficiency and iron overload, and leads to improved outcomes.

Marwa Zakaria
Associate Professor of Pediatrics,
Faculty of Medicine,
Pediatric Department,
Zagazig University,
Zagazig, Egypt

Tamer Hassan
Professor of Pediatrics,
Faculty of Medicine,
Pediatric Department,
Zagazig University,
Zagazig, Egypt

Section 1

Iron from Food to Consume

Chapter 1

Dietary Iron

Kouser Firdose and Noor Firdose

Abstract

Iron metabolism differs from the metabolism of other metals in that there is no physiologic mechanism for iron excretion, it is unusual; approximately 90% of daily iron needs are obtained from an endogenous source, the breakdown of circulating RBCs. Additionally humans derive iron from their everyday diet, predominantly from plant foods and the rest from foods of animal origin. Iron is found in food as either haem or non-haem iron. Iron bioavailability has been estimated to be in the range of 14–18% for mixed diets and 5–12% for vegetarian diets in subjects with no iron stores. Iron absorption in humans is dependent on physiological requirements, but may be restricted by the quantity and availability of iron in the diet. Bioavailability of food iron is strongly influenced by enhancers and inhibitors in the diet. Iron absorption can vary from 1 to 40%. A range of iron bioavailability factors that depend on the consumption of meat, fruit, vegetables, processed foods, iron-fortified foods, and the prevalence of obesity. The methods of food preparation and processing influence the bioavailability of iron. Cooking, fermentation, or germination can, by thermal or enzymatic action, reduce the phytic acid and the hexa- and penta-inositol phosphate content. Thus improving bioavailability of non-haem iron. This chapter will elaborate the dietary iron sources and means of enhancing bioavailability.

Keywords: iron, diet, haem iron, non haem iron, bioavailibility, fortification, biofortification

1. Introduction

Iron has an essential physiologic role, as it is involved in oxygen transportation and energy formation. The body cannot synthesize iron and must acquire it. Though the human body recycles and reutilizes iron, it also loses some iron daily; these lost pools require replacement. However recycling the iron from senescent erythrocytes meets most of the body's iron needs by macrophages; only 5–10% of iron requirements come from food [1].

Iron differs from other minerals because iron balance in the human body is regulated by absorption only and there is no physiologic mechanism for excretion [2].

Haem iron derived from animal sources is better absorbed than non-haem iron derived from plant sources, whole cereals, whole pulses, and vegetables, particularly green leafy vegetables, contribute to a significant intake of dietary iron [3].

IntechOpen

Dietary iron bioavailability depends primarily on the availability of iron for absorption in the GI tract, determined by the physicochemical form of iron in the lumen of the GI tract, largely dictated by the composition of meals, and secondarily by an individual's absorptive efficiency, which depends on physiological require-ments for iron and homeostatic mechanisms designed to maintain null balance. Bioavailability factors have been derived based on the balance of enhancers and inhibitors of iron absorption in diet [4].

Various strategies can be adopted to enhance bioavalibility and to combat iron deficiency which includes dietary diversification, food fortification, weekly iron and folic acid supplementation among others [1].

Iron is found naturally in many foods and is added to some fortified food prod-ucts; recommended amounts of iron can be obtained by eating a variety of foods, including non-vegetarian food viz. lean meat, seafood, and poultry etc. in addition to the iron-fortified breakfast cereals and breads, white beans, lentils, spinach, kidney beans, and peas, nuts and some dried fruits [5].

Food diversification approach designed to increase micronutrient intake through diet represents the most desirable and sustainable method for preventing iron deficiency [3].

Reference intakes are used for a wide range of activities, such as planning diets, formulating complementary foods, setting levels of food fortification, implementing biofortification programs, and food labeling [4].

2. Types of dietary iron

Dietary iron has two primary forms: haem and non-haem [1, 2, 6]. Haem iron has a higher bioavailability and is absorbed easier without the need for absorption-enhancing cofactors (**Figure 1**) [1, 2].

Haem iron is estimated to contribute 10–15% of total iron intake in meat-eating populations, but, because of its higher and more uniform absorption (estimated at 15–35%), it could contribute 40% of total absorbed iron during iron deficiency to about 10 percent during iron repletion [7].

Non-haem iron, which is the most important dietary source in vegetarians, shows lower bioavailability [1, 2]; All non-haem food iron that enters the com-mon iron pool in the digestive tract, however, it is important to note that not all fortification iron enters the common pool [2]; 17% of dietary non-haem iron gets absorbed [1].

Studies shows that, iron bioavailability is estimated to be 14–18% for mixed diet consumers and 5–12% for vegetarian diet consumers. Thereby, less than one-fifth of the dietary iron gets absorbed by the body [1].

Iron absorption in humans is dependent on physiological requirements, but may be restricted by the quantity and availability of iron in the diet [8]. Body absorbs iron from plant sources better when eaten with meat, poultry, seafood, and foods that contain vitamin C, like citrus fruits, strawberries, sweet peppers, tomatoes, and broccoli [5].

The diets of omnivores contain relatively small quantities of haem iron derived from meat and fish, which is always well absorbed [7, 8]. The remainder of the soluble iron forms a common non-haem iron pool and absorption is very variable, depending on meal composition, but its absorption is strongly controlled by iron stores [8].

Heme Iron	Non-heme Iron
Animal food source	Plant food source
Part of heme protien attached to the iron	Doesnot contain a of heme protien attached to the iron
Occurs in oysters, red meat, poultry, beef liver and fish	occurs in beans, nuts, lentils green leafy vegetables
Absorption rate is high	Not well absorbed as heme iron
Excess heme iron can cause health risk	Doesnot cause health risks

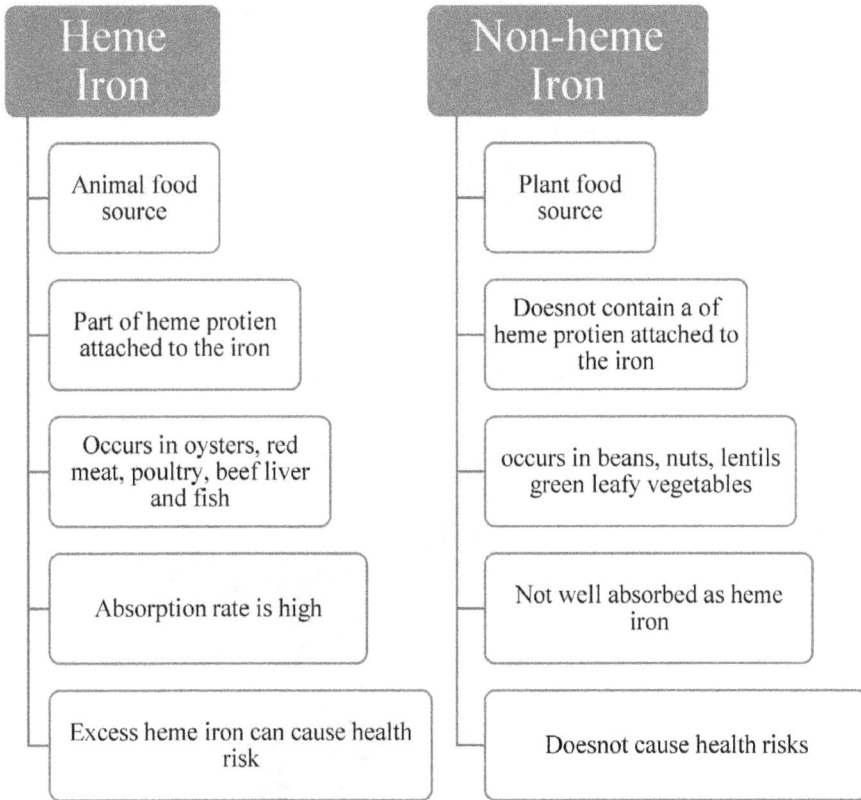

Figure 1.
Types of iron.

3. Sources of iron

3.1 Non-haem sources of iron

High nutrition benefits of coarse cereals point to the need for an increase in their consumption and even higher production (**Figure 1a**).

Top five pulses with respect to iron content are horse gram dal, soybean, moth beans, lentil (whole), and Bengal gram (whole). Horse gram dal, soybean, and moth beans provide as much as twice the iron in comparison to green gram dal and red gram dal. Arhar and moong, though lowest in iron content.

Bajra, ragi, rice flakes (poha) wheat flour, and jowar provide a higher amount of iron than maize and rice. Rice has the lowest iron content.

Green leafy vegetables are considered to be rich sources of iron and calcium. For example, beet greens, pumpkin leaves, colocasia leaves, and radish leaves having very high-iron content is usually not consumed by people and rather discarded as waste. There are others such as curry (8.7 mg/100 g), mint (8.6 mg), parsley (5.5 mg), coriander (5.5 mg), and drumstick (4.6 mg) though high in iron content, are consumed either less frequently or in small quantities. Greens like spinach, mustard leaves, and bathua leaves though popular are those with the least iron content [3].

Some of the nuts and oilseeds such as gingelly seeds (14.9 mg), mustard seeds (13.5 mg), cashew nuts (5.9 mg) and almond (4.5 mg), are fairly rich sources of dietary iron. Most of the fruits and vegetables, except lotus stem (3.3 mg iron/100 g), are a poor source of iron.

Jaggery, though rich in iron (4.6 mg iron per 100 g), is usually consumed in small amounts. Promoting traditional Indian snacks like gur chana or tilbugga prepared from jaggery and Bengal gram or gingelly seeds can contribute to significantly higher intake of jaggery and thus iron [5, 7].

3.2 Haem sources of iron

Among poultry, chicken liver is the richest source of iron (9.9 mg/100 g) followed by duck meat (4.3 mg/100 g) (**Figure 2b**).

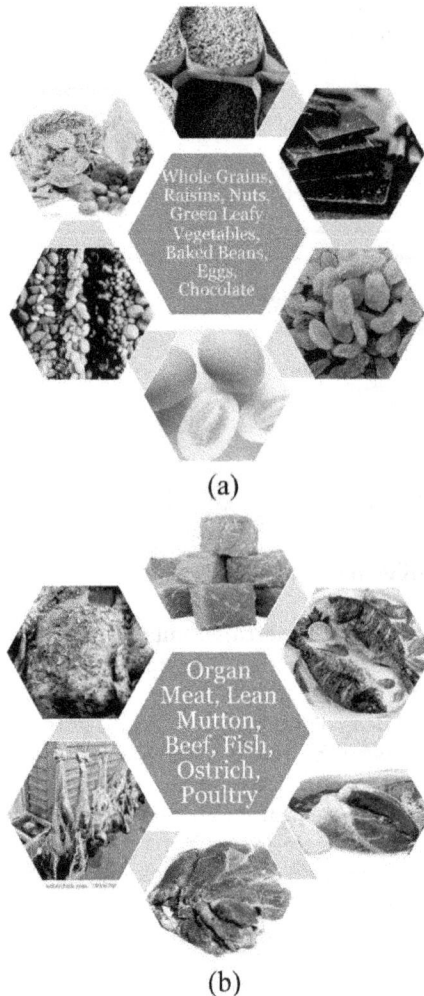

(a)

(b)

Figure 2.
a. Sources of non-heme iron. b. Sources of heme iron.

Animal meat, particularly liver and spleen, is very rich source of iron.
Boiled egg yolk is rich in iron as compared to egg white.
Fish on the contrary are not a very good source of iron [3].

Iron is present in a variety of foods, so eating a varied and healthful diet is important. Since Vitamin C enhances the absorption of iron, eating iron rich foods along with a source of vitamin C (citrus fruits and juices, etc.) can help replenish your body's iron stores. Nevertheless, iron may be absorbed into foods that have been cooked in iron cookware [9].

Common sources of Iron are depicted in **Table 1**.

- Iron contamination: For cooking, sometimes an iron skillet is a utensil used for cooking vegetables and other food to increase iron content in that food. Such a source of contaminated iron is sometimes practiced in some regions of the world [1].

Common Sources of Vitamin C are depicted in **Table 2**.

Victuals	Portion size (approx.)	Amount of iron
Haem sources		
Beef liver	85 g	5.2 mg
Beef-ground	85 g	2.2 mg
Canned clams	85 g	23.8 mg
Chicken breast	85 g	1.1 mg
Chicken liver	85 g	10.8 mg
Fish, tuna canned	85 g	1.3 mg
Lamb	85 g	3.0 mg
Large egg	1	1.0 mg
Oysters	85 g	13.2 mg
Pork	85 g	2.7 mg
Sirloin streak	250 g	1.6 mg
Shrimp	85 g	2.6 mg
Salmon	100 g	1.28 mg
Tofu	100 g	8 mg
Turkey, dark meat	85 g	2.0 mg
Turkey, light meat	85 g	1.1 mg
Non haem sources		
Greens/veggies:		
Beets, canned	64 g	1.5 mg
Brussel sprouts	64 g	2.0 mg
Collards or beet	64 g	1.2 mg
Dried thyme	5 g	1.2 mg
Greens	125 g	2.2 mg
Mushrooms	64 g	1.4 mg

Victuals	Portion size (approx.)	Amount of iron
Peas, frozen	64 g	1.2 mg
Potato, baked with skin on	Medium size	1.9 mg
Swiss chard	64 g	2.0 mg
Spinach cooked/raw	64 g/128 g	3.0 mg
Sweet potato, baked with skin on	Medium size	1.1 mg
Sauerkraut, canned	64 g	1.7 mg
Tomato Sauce	64 g	1.3 mg
Nuts		
Almonds or pistachios	32 g	1.3 mg
Walnuts	85 g	1.0 mg
Dried peaches	64 g	1.6 mg
Dried raisins	64 g	1.4 mg
Dried plums	64 g	1.3 mg
Dried apricots	64 g	1.2 mg
Pine or cashews	85 g	1.6 mg
Prune juice	125 g	3.2 mg
Strawberries	1 pint	1.5 mg
Beans:		
White	100 g	5.8 g
(Black, pinto)	64 g	1.6–1.8 mg
(Kidney, lima)	64 g	2.6–3.9 mg
Soybeans	64 g	4.4 mg
Tofu, firm	64 g	3.4 mg
Chickpeas	100 g	2.4 mg
Double beans (cooked)	125 g	4.5 mg
Tomato (sun dried)	125 g	4.9 mg
Soy milk	300 ml	2.7 mg
Quinoa	125 g	2.8 mg
Kale	125 g	1.1 mg
Grains:		
Lentils	64 g	3.5 mg
Pumpkin seeds	28 g	4.2 mg
Cereal	64 g	2–12 mg
Cream of wheat	64 g	5.2 mg
Oat meal	64 g	1.7 mg
Oatmeal Instant fortified with iron	64 g	5.0 mg

Table 1.
Sources of iron [9, 10].

Fruits	Vegetables
Amla,	Amaranthus
Cashew fruit,	Agathi
Guava,	Brussels
Lakuch,	Carrot
Korukkapalli,	Coriander
Papaya,	Cabbage
Lime, sweet (Malta)	Drumstick,
Musambi,	Fetid cassia,
Lemon,	Knol-khol radish,
Muskmelon	Turnip,
Orange	Parsley
Pineapple	*Sprouts are richer source of ascorbic acid.
Ripe tomato	
Zizypus	

Table 2.
Sources of vitamin C [7].

4. Bioavailability: dietary iron absorption

The bioavailability of dietary iron is the proportion of iron that is actually available for absorption and utilization by the body (**Figure 3**) [11]. In humans, haem iron is well absorbed and its absorption varies little with the composition of the meal. Absorption is inversely related to the quantity of iron stores in the body [6].

4.1 Factors influencing dietary iron absorpion

4.1.1 Haem iron absorption

- Iron status of subject: absorption ranges from 15 to 25 percent in normal subjects and 25–35 percent in iron-deficient subjects.

- Amount of dietary haem iron, especially from meat

- Content of calcium in meal (e.g. milk, cheese)

- Food preparation (time, temperature): Baking and prolonged frying have been shown to reduce haem iron absorption by about 40 percent.

4.1.2 Non-haem iron absorption

- Iron status of subject: The absorption of non-haem iron ranges from 2 to 20 percent. Severely iron-deficient individuals absorb non-haem iron at higher rates than those with normal iron levels. Absorption was shown to be the highest (5–13 percent) in pregnant anemic women.

- Concomitant diet: The specific rate of absorption of non-haem iron is highly dependent on the effect of concomitantly ingested dietary components (reducing substances such as ascorbic acid keep iron in the reduced ferrous form) and the amount of body iron stores.

Iron bioavailability

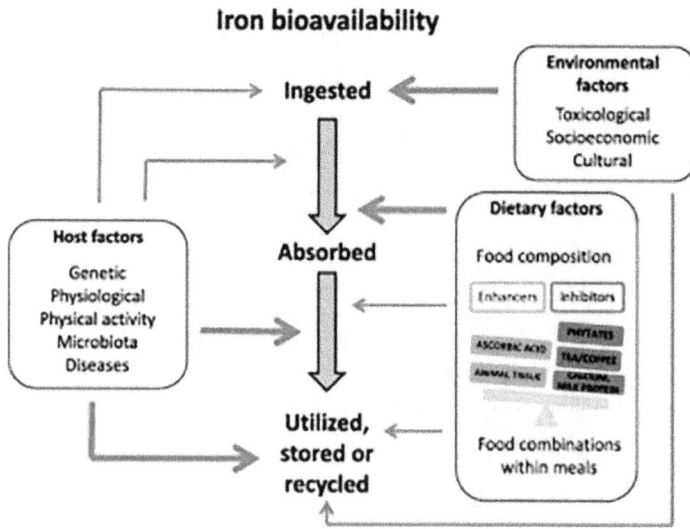

Figure 3.
Iron bioavailability.

- Food preparation (time, temperature): Cooking of cereals and pulses was shown to cause a loss of 22–24 percent of their iron, however, baking chapatti on an iron plate raised the iron content by 19 percent.

- Fermentation can degrade the phytate and increase the bioavailability of iron in bread made from whole wheat flour.

- Household processes such as germination, malting of grains/pulses and fermentation should be used to over come phytates and enhance the ascorbic acid and B-vitamins.

- Amount of potentially available non-haem iron (adjustment for fortification iron and contamination iron).

4.2 Factors effecting bioavailibility

Many different dietary components either enhance or inhibit dietary iron absorption when they are simultaneously present in the diet [1]. The bioavailability of food and dietary iron is influenced by certain factors, some of which are briefly described below [6].

4.2.1 Enhancers

Iron Absorption Enhancers foods are those, when you eat them together with a natural source of iron or an iron supplement, they help aid body's ability to effectively absorb the iron into body system [1]. There's no point in ingesting iron if the body cannot absorb it [10].

1. Meat/fish/poultry—these are also sources of the most potent form of iron (haem iron)

2. Acidic fruits—oranges/orange juice /cantaloupe/strawberries/grape fruit etc.

3. Vegetables—broccoli/brussels, sprouts/tomatoes/tomato Juice/potatoes/green and red peppers etc. [6, 10]

4. Fermented vegetables (e.g. sauerkraut), fermented soy sauces, etc. [6]

4.2.1.1 Mechanism

1. MFP Factor: It is a peptide present in meat, fish, and poultry. It enhances the absorption of non-haem iron present in the same meal. The detailed underlying mechanism is still not known. However, evidence suggests that cysteine-containing peptides present in the meat act by inhibiting luminal inhibitors and eventually form luminal carriers for iron transportation.

   • Studies consistently showed an enhanced effect on vegetarian iron absorption by 2-3-fold by adding animal proteins.

2. Ascorbic Acid (Vit C): This effect is mainly due to its iron-chelating and reducing abilities, converting ferric iron to ferrous iron, which has higher solubility and better absorption by 75–98 percent. The addition of ascorbic acid to cereals and pulses enhanced the available iron. [6] Vitamin C also has been shown to have an inhibitory effect on iron absorption inhibitors such as phytate, polyphenols, and calcium. Studies have convincingly shown the dose-dependent enhancing effect of natively present or added vitamin C on iron absorption [1].

   • The comprehensive review has shown that a food source containing 50 mg of ascorbic acid consumed with the main meal providing most of the daily intake of iron enhances iron bioavailability significantly.

Examples
Meat, fish, poultry

• The addition of 90–100 g of meat, fish or poultry to the daily diet, significantly improves the bioavailability of iron. Meat and fish taken even in small amounts markedly improve the bioavailability of non-haem iron.

• A non-vegetarian diet containing 3 oz. (approximately 85 g) of meat provides the same increase in non-haem iron absorption as 75 mg of ascorbic acid

• Eggs are rich in iron content, but its bioavailability is poor. However, as a source of iron, eggs should be eaten along with a fruit or any other source containing 100 mg of ascorbic acid, or between meals [6].

Vitamin C

• In cereal-based diets, absorption was the best for rice and vegetable combinations, which may result from ascorbic acid present in the vegetables. Children who consumed GLV once a week or more frequently had higher iron levels than non-consumers.

- Daily intake of guava fruit with the two major meals by young anemic women shows significant increase in iron.

- In regional meals, the addition of citrus fruit juices or a portion of potato, cauliflower or cabbage increases iron availability markedly [6].

- If 25 mg of ascorbic acid as lemonade is consumed at two meals a day, it doubles the absorption of iron from a meal and improves the iron status [6].

- The enhancing effect of ascorbic acid is dose-dependent, but little extra benefit is derived by increasing the intake of ascorbic acid beyond 100 mg in a meal. The influence of ascorbic acid is greatest on meals with low iron bioavailability, such as vegetarian meals [6]. It also improves the availability of iron from fortified foods.

4.2.2 Inhibitors

The following are Iron Absorption Inhibitors. i.e. when you have them together with a source of iron, they will either inhibit (limit) or prevent your body from absorbing the iron, you ingested, these foods should be avoided when taking iron rich foods in diet. This also includes any supplementations.

1. Coffee and tea [6, 10] cocoa, certain spices, certain vegetables and most red wines. (Iron-binding phenolic compounds) [6]

2. Vegetables—spinach*/chard/beet greens/rhubarb/sweet potatoes whole grains and bran [6, 10].

3. Bread made from high-extraction flour, breakfast cereals, oats, rice [especially unpolished rice], pasta products, cocoa, nuts, soybeans and peas

4. Calcium (e.g. milk, cheese) [6].

5. Isolated soy ingredients—products made with soy flour and isolated soy protein concentrate [6, 10].

4.2.2.1 Mechanism

- Phytates: they are known inhibitors of non-haem iron absorption [10]. Food sources high in phytates include soybean, black beans, lentils, mung beans, and split beans. Unrefined rice and grains also contain phytate [1]. Phytates can decrease non-haem iron absorption by 51–82 percent, and are found in higher concentrations in unrefined, non- or under-milled cereals than in refined, milled cereals [6].

- Polyphenols: they are commonly found in tea as tannic acid and also in red wine and oregano. They inhibit non-haem iron by binding within the intestine [1, 6, 10].

- Calcium: calcium has been found to have an inhibitory effect on both haem and non-haem iron absorption. Its exact mechanism is unclear [1, 6]. The first 40 mg of calcium in a meal showed no inhibiting effect, whereas 300–600 mg of calcium inhibited iron absorption by 60 percent, which is the maximum inhibition of iron. Studies showed that about 30–50 percent more iron was absorbed when no milk or cheese was served with the main meal, which provided most of the dietary iron [6].

Examples

- Approximately 250 ml of black tea can inhibit non-haem iron absorption by approximately 50 percent even when drunk 1 hour after consuming the meal; however, it has no effect when consumed between meals. This inhibition is strongly dose-related, which can be reduced to some extent by serving tea with lemon or adding sufficient milk (100 ml) to the cup of tea [6].

- Iron absorption is affected less by coffee than tea.

- To overcome the inhibitory effects, tea or coffee should not be consumed with the main iron-containing meals [6].

- Milk is better to be avoided with the main meals that contribute most of the daily iron intake, however it can be taken at breakfast, in the evening or at bedtime. Milk intake may be increased to as much as 400 ml per day provided it is distributed as suggested.

- The high iron availability of breast milk, which averages 50 percent (compared to 10–20 percent in cow's milk), is reduced when breast milk is taken together with cow's milk or weaning foods. Hence weaning foods are recommended to be given separately from the breast milk [6].

- Spinach is a good source of iron, too, but it is best to cook the spinach first—it unlocks much of the iron potential for it.

- Practical solutions for the competition of calcium with iron is to increase iron intake, increase its bioavailability or avoid taking calcium and iron-rich foods at the same time [6].

- The presence of carotene in rice-, wheat- and corn-based diets improved iron absorption from one to more than threefold suggesting that both ascorbic acid and carotene prevented the inhibitory effect of phytates on iron absorption [6].

Dietary factors that influence iron absorption, i.e. enhancers and inhibitors, have been shown repeatedly to influence iron absorption in single-meal isotope studies, whereas in multimeal studies with a varied dietary factor, the effect of single components have been, as expected, more modest [2].

The iron status of the individual and other host factors, such as obesity [2] and medical problems like malabsorptive disorder, Celiac disease, Crohn's disease and those with history of gastric bypass surgery interferes with iron absorption, play a key role in iron bioavailability, and iron status generally has a greater effect than diet composition. Hence to develop a range of iron bioavailability factors based not only on diet composition but also on subject characteristics, such as iron status and prevalence of obesity is the need of the time [1].

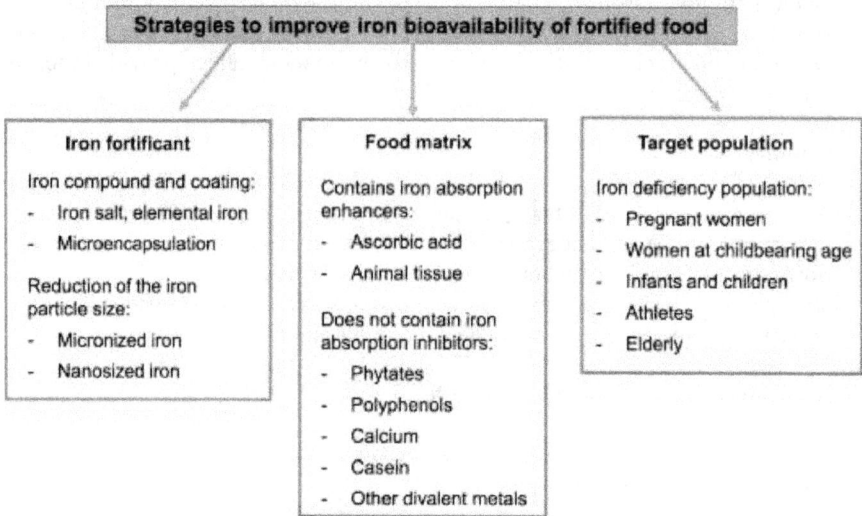

Figure 4.
Strategies to improve iron bioavailability.

The bioavailability of iron differs in various food sources depending on the types of dietary iron and the presence or absence of iron absorption enhancers or inhibitors among others (**Figure 4**) [2, 11].

5. Fortification

Food fortification is the addition of micronutrients at the point of manufacture to enhance the nutritional content of the food items, such as meal ingredients or condiments [12].

Fortification is a medium-to-long-term approach that requires a suitable food vehicle and organized processing facilities. About 34 current evidence indicates that food fortification is an effective and cost-effective strategy for reducing the prevalence of iron deficiency. [8, 13, 14] providing extremely good value, with its benefits far outweighing the costs [13] in populations that consume diets containing suboptimal quantities of bioavailable iron [9] and WHO/FAO recommends that the level of fortification is based on the estimated daily iron intake deficit adjusted for bioavailability [8].

Since iron-deficiency anemia is a main indicator of micronutrient deficiencies, one of the safest strategies available to reduce the risk of iron deficiency is fortification with low doses of iron homogeneously diluted in a larger mass of food remains. These considerations are important in the context of the United Nations Sustainable Development Goals, alongside the several servings of iron per day. Unlike supplementation, iron fortification at the point of manufacture enables the delivery of small doses of the micronutrient in a food vehicle. It is slower to raise body iron levels compared with iron supplementation or iron therapy, but it might be safer [8].

Iron fortification can be done through staple food items such as rice, oils, and wheat; condiments such as fish sauce, soy sauce, lentils (**Figure 5**) [15], salt and sugar; and lastly through processed commercial food items, including infant complementary foods, dairy products, and noodles.

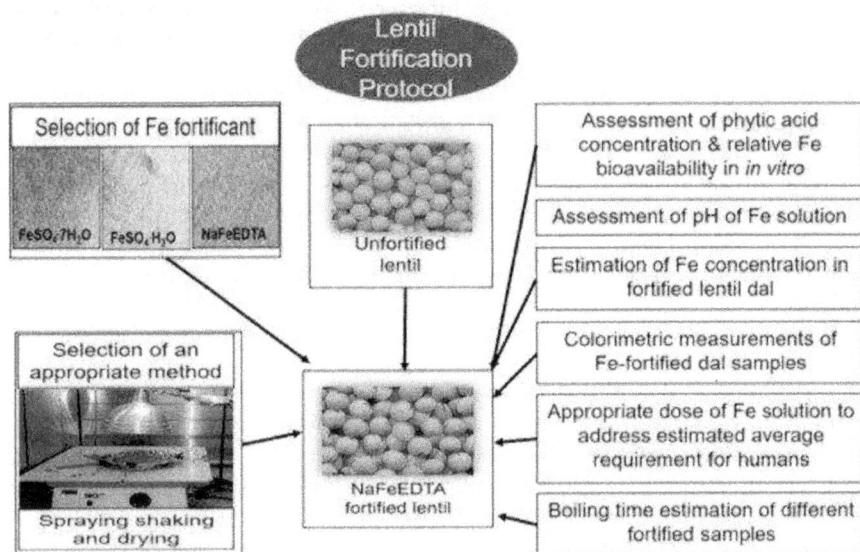

Figure 5.
Lentil iron fortification.

Compared to supplements, the use of fortified complementary foods has been shown to be safer and more effective, since the limitations associated with supplementation includes the need to purchase iron supplements and the need for a higher degree of treatment compliance [13]. The high compliance to fortification is due to the ease of substitution of unfortified staples with fortified foods 34.

Iron salts recommended by WHO for fortification include ferrous sulphate, ferrous fumarate, ferric pyrophosphate, and electrolytic iron powder [2, 13].

The WHO drew up Guidelines for food fortification which included fortification with iron [14]. In a recent directive, the WHO and partner organizations, while providing guidance on national fortification of wheat and maize flours, have endorsed NaFeEDTA to be the only Fe fortificant suitable for use in high-extraction flours [16].

Wheat is currently the primary staple food for nearly one-third of the world's population. NaFeEDTA protects iron from the phytic acid present in foods by binding more strongly to ferric Fe at the pH of the gastric juice in the stomach and then exchanging the ferric Fe for other metals in the duodenum as the pH rises [12]. It is 2- to 4-fold more bioavailable than ferrous sulfate, particularly in meals with a high-phytate content, thereby making it ideal for use in whole wheat flour [16].

Many research studies were undertaken globally on food fortification with iron; with wheat flour fortification the evidence for reducing iron deficiency among women in reproductive age (WRA) is consistent but on reducing anemia is limited [14].

The three reviews wherein multiple vehicles and various iron sources including electrolytic iron have been used concluded that consumption of iron fortified foods results in:

1. Improvement in weighted mean difference (WMD) in Hb of 0.42 g/dL, increase in serum ferritin of 1.37 µg/L and reduced risk of being anemic and iron deficient in children;

2. Improvement in standardized mean difference (SMD) in Hb of 0.55 and 0.64 g/dL, serum ferritin of 0.91 and 0.41 µg/L and reduced risk of being anemic RR 0.55 and 0.68 in children <15 8 years and WRA, respectively;

3. Improvement in WMD in Hb of 0.51 g/dL in children <10 years [14].

Efficacy of NaFeEDTA, as a fortificant has also been demonstrated in food vehicles such as curry powder, sugar, fish sauce, and maize flour [11, 13, 14, 16].

Salt: The National Institute of Nutrition had developed a technology for fortification of salt with iron and extensively tested its safety and efficacy. Fortification standards were formulated to provide 1 mg of iron (and 15 µg of iodine) per gram of salt which provides about 30–60% of RDA of 17 mg of an adult man consuming 5–10 g salt per day (FSSAI) [14].

Fortification of salt with iron is preferred because it requires only a relatively small volume of the food stuff to be fortified, unlike fortification of cereals. Currently iodisation of salt is nearly universal and using this platform it will be possible to scale up production, distribution and marketing of DFS. Double fortified salt-Iron fortified Iodized salt (providing about 10 mg of iron/day).

The studies on impact of fortified salt with three types of technologies ($FeSO_4$, 13 encapsulated ferrous fumarate and ferric pyrophosphate) showed;

1. improvement in SMD of Hb of 0.44 g/dL and ferritin 0.62 µg/L,

2. anemia risk reduction ratio of 0.16 and IDA 0.20 [14].

Since haem iron is readily bioavailable, there have been some instances of the use of meat-derived products in packaged food as fortificants [12].

Iron fortification is not a standalone strategy to correct iron deficiency. There is a need to improve dietary diversification especially consumption of vitamin C rich fruits along with meals so that iron bioavailability is improved [14]. Point-of-use fortification employs micronutrient powders in the form of packed, single-dose sachets that can be added to prepared food to improve its nutrient value [12].

5.1 Benefits of iron fortification

Food fortification offers many health benefits.

1. Iron fortification in children led to improvement in iron and hemoglobin status.

2. Hemoglobin levels significantly increased by 6.2 g/L and the risk of anemia was 50% lower in children receiving fortified milk or infant cereals [13]. Use of fortified milk and cereal-based products are more effective in reducing anemia in young children in developing countries, compared to the use of non-fortified products.

3. Cereal flour fortification with Fe is the most cost-effective and sustainable way to improve its status in deficient populations [16].

Food-fortification practices vary nationally and the need to adjust the dietary iron bioavailability factor for fortification iron will depend on the proportion of fortification iron in the total iron intake and the iron compounds used [13].

Iron compounds used for the fortification of foods will only be partially available for absorption. Once iron is dissolved, its absorption from fortificants and food contaminants is influenced by the same dietary factors [7].

Bioavailability of fortification iron varies widely with the iron compound used, and foods sensitive to color and flavor changes are usually fortified with water-insoluble iron compounds of low bioavailability [2].

6. Biofortification: iron bio-fortified crops

Biofortification involves the targeted breeding of staple food crops in order to increase their intrinsic content of micronutrients, including iron. By combining traditional breeding with modern techniques, biofortification blends the traits of high-yield crop varieties with high iron varieties (**Figure 6**) [12, 17].

The levels of iron for wheat and rice fortification is similar and permit additions ranging from a minimum of 33% to a maximum of 100% of RDA of 17 mg [14].

The use of biofortified crops address micronutrient deficiencies by enriching the staple food items that constitute the main portion of the diet. Iron biofortification is applicable to cereals like wheat, rice, and millet [12, 14] and to pulses like beans, peas, and lentils [12].

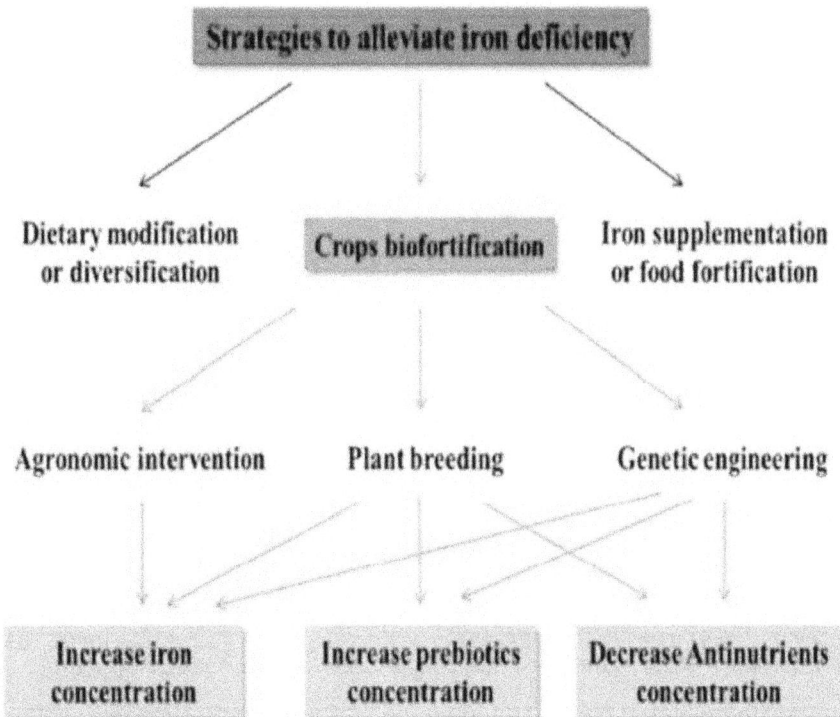

Figure 6.
Biofortification.

Therefore, even very small amounts of micronutrients could have a positive impact over time. Secondly, if biofortified crops also possess excellent agronomic characteristics, a self-sustaining public health intervention will result because farmers will favor such crops.

Iron-biofortified millet contains higher concentrations of iron. Iron levels in this type of millet reaches 90 ppm, whereas levels in nonbiofortified millet are around 20 ppm. Several studies indicate that regular intake of biofortified millet can be efficacious against iron deficiency [12].

A pearl millet variety was studied among 12–16 year adolescent girls consuming 200–300 g of pearl millet during lunch and dinner for 4 months revealed the following:

a. There was no difference in Hb

b. Ferritin increased significantly and

c. Positive impact on cognitive function [14].

Biofortified pulses, containing 100 ppm or more, have the highest concentrations of iron. Several studies have examined the bioavailability or efficacy of iron in biofortified beans consumed in developing countries; though phytic acid is present in beans, a high proportion of the iron is contained in phytoferritin. Iron from ferritins has been shown to be highly bioavailable [12].

7. Conclusion

The overall intake of iron from iron rich foods together with Vitamin C needs to be increased to obtain the optimum level of recommended dietary allowance of iron. This increase should be merged with efforts to cartel appropriate foods in the diet to enhance the bioavailability of iron and reduce inhibitory factors. Even without the haem iron found in fish or poultry, vegetarians are not at greater risk from iron deficiency than non-vegetarians. Cereals and millets, pulses and legumes, Green Leafy Vegetables, nuts and oilseed are good sources of iron.

The food combinations should be designed on the basis of foods that are normally consumed, accustomed, locally available and low-cost; comprising enhancing factors and limiting inhibitors to the extent possible and providing an overall balanced diet to provide all the major nutrients required by the body. In addition, combinations and proportions of foods on the basis of the factors influencing dietary iron absorption, a balanced diet has to be calculated.

Dietary consumption of iron and ascorbic acid could be increased by encouraging the production, processing, marketing and consumption of foods rich in these nutrients. Vitamin C-rich foods must be consumed at the same meal that contributes the major part of daily dietary iron.

Nutrition education could be a means for further promotion of dietary iron.

Conflict of interest

"The authors declare no conflict of interest."

Dietary Iron
DOI: http://dx.doi.org/10.5772/intechopen.101265

Author details

Kouser Firdose[1]* and Noor Firdose[2]

1 Department of Ilmul Qabalat wa Amraze Niswan, National Institute of Unani Medicine, Bangaluru, India

2 Family Physician, Health Care Centre, Bengaluru, India

*Address all correspondence to: kouser2fathima@gmail.com

IntechOpen

References

[1] Moustarah F, Mohiuddin SS. Dietary Iron. Treasure Island (FL): StatPearls Publishing; 2021

[2] Hurrell R, Egli I. Iron bioavailability and dietary reference values. The American Journal of Clinical Nutrition. 2010;**91**(5):1461S-1467S

[3] Taneja DK, Rai SK, Yadav K. Evaluation of promotion of iron-rich foods for the prevention of nutritional anemia in India. Indian Journal of Public Health. 2020;**64**:236-241

[4] Fairweather-Tait S, Speich C, Mitchikpè CE, Dainty JR. Dietary iron bioavailability: A simple model that can be used to derive country-specific values for adult men and women. Food and Nutrition Bulletin. 2020;**41**(1):121-130

[5] NIH. Iron Fact Sheet for Consumers [Internet]. 2019. Available from: https://ods.od.nih.gov/factsheets/Iron-Consumer/pdf [Accessed: July 17, 2021]

[6] Sharma KK. Improving bioavailability of iron in Indian diets through food-based approaches for the control of iron deficiency anaemia. Food Nutrition and Agriculture. 2003;**32**:51-61

[7] World Health Organization. Human Vitamin and Mineral Requirements. Report of a Joint FAO/WHO Expert Consultation. Bangkok, Thailand: Food and Agriculture Organization of the United Nations. 2001:195-216

[8] Dainty JR, Berry R, Lynch SR, Harvey LJ, Fairweather-Tait SJ. Estimation of dietary iron bioavailability from food iron intake and iron status. PLoS One. 2014;**9**(10):e111824

[9] Cedars-Sinai. Iron Rich Foods [Internet]. 2018. Available from: https://www.cedars-sinai.org/content/dam/cedars-sinai/programs-and-services/blood-donor/documents/iron-rich-foods.pdf [Accessed: June 26, 2021]

[10] Iron Rich Foods. [Internet]. 2018. Available from: http://edinaschools.org/cms/lib07/MN01909547/Centricity/Domain/304/Iron_Guide.pdf [Accessed: August 02, 2021]

[11] Blanco-Rojo R, Pilar Vaquero M. Iron bioavailability from food fortification to precision nutrition. A review. Innovative Food Science & Emerging Technologies. 2018;**51**:126-138. DOI: 10.1016/j.ifset.2018.04.015

[12] Prentice AM, Mendoza YA, Pereira D, Cerami C, Wegmuller R, Constable A, et al. Dietary strategies for improving iron status: Balancing safety and efficacy. Nutrition Reviews. 2017;**75**(1):49-60

[13] Dobe M, Garg P, Bhalla G. Fortification as an effective strategy to bridge iron gaps during complementary feeding. Clinical Epidemiology and Global Health. 2018;**6**(4):168-171

[14] Nair KM. Contextualizing the principles of iron fortification of foods in India. Bulletin of the Nutrition Foundation of India. 2019;**40**(2):1-5

[15] Podder R, Tar'an B, Tyler RT, Henry CJ, DellaValle DM, Vandenberg A. Iron fortification of lentil (lens culinaris medik.) to address iron deficiency. Nutrients. 2017;**9**(8):863. DOI: 10.3390/nu9080863

[16] Muthayya S, Thankachan P, Hirve S, Amalrajan V, Thomas T, Lubree H, et al.

Dietary Iron
DOI: http://dx.doi.org/10.5772/intechopen.101265

Iron fortification of whole wheat flour reduces iron deficiency and iron deficiency anemia and increases body iron stores in Indian school-aged children. The Journal of Nutrition. 2012;**142**(11):1997-2003

[17] Yin X, Yuan L. Biofortification to struggle against iron deficiency. In: Phytoremediation and Biofortification. Springer Briefs in Molecular Science. Dordrecht: Springer; 2012. pp. 59-74. DOI: 10.1007/978-94-007-1439-7_4

Section 2

Iron Metabolism

Chapter 2

Iron in Cell Metabolism and Disease

Eeka Prabhakar

Abstract

Iron is the trace element. We get the iron from the dietary sources. The enterocytes lining the upper duodenal of the intestine absorb the dietary iron through a divalent metal transporter (DMT1). The absorbed ferrous iron is oxidized to ferric iron in the body. This ferric iron from the blood is carried to different tissues by an iron transporting protein, transferrin. The cells in the tissues take up this ferric form of iron by internalizing the apo transferrin with its receptors on them. The apo transferrin complex in the cells get dissociated resulting in the free iron in cell which is utilized for cellular purposes or stored in the bound form to an iron storage protein, ferritin. The physiological levels of iron are critical for the normal physiology and pathological outcomes, hence the iron I rightly called as double-edged sword. This chapter on iron introduces the readers basic information of iron, cellular uptake, metabolism, and its role cellular physiology and provides the readers with the scope and importance of research on iron that hold the great benefit for health care and personalized medicine or diseases specific treatment strategies, blood transfusions and considerations.

Keywords: iron, micronutrient, cell absorption, homeostasis, iron deficiency, iron overload, diseases

1. Introduction

Iron is a chemical element with symbol Fe and atomic number 26 (**Figure 1**). Classified as a transition metal, iron is a solid at room temperature. The symbol "Fe" is derived from the Latin *ferrum* for "firmness", iron exists in many different forms in nature. Iron makes up 5% and the chief constituent of earth's crust. It is the second abundant metal on earth and is most abundant as an alloy. Although its most abundance, it is required in very small amounts and hence is also called a trace element. The readers should also appreciate the meaning of "trace" here, in other sense that its trace amounts are very critical, and any more amounts of this element is as dangerous as its normal functions in the cell. Hence, the iron is called to be a double-edged sword [1]. Iron is a metal which belong to the transition metals group or VIIIB elements group of periodic table (**Figure 2**). Iron exists in different forms in nature [2]. The pure metal is very reactive chemically and rapidly corrodes, especially

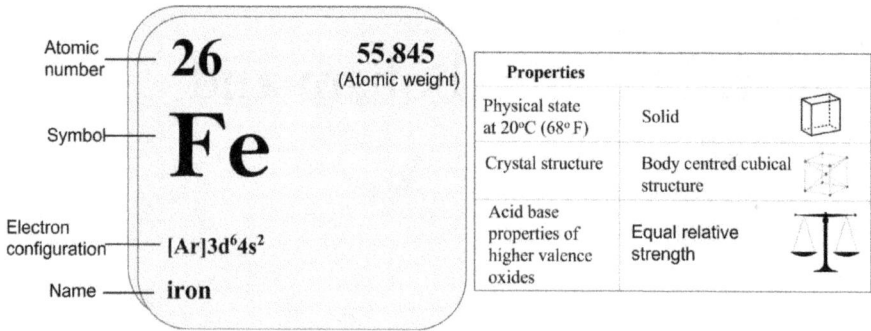

Figure 1.
Properties of iron (source: adapted from Encyclopædia Britannica, *Inc).*

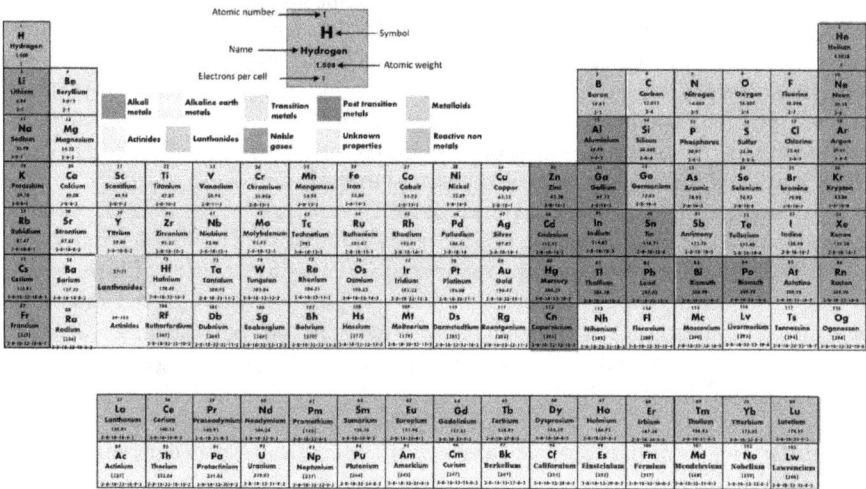

Figure 2.
Periodic table of elements representing element properties and their categories based on the similarities in physicochemical properties and atomic number.

in moist air or at elevated temperatures. It has four allotropic forms or ferrites, known as alpha, beta, gamma, and omega, with transition points at 700, 928, and 1530C. The alpha form is magnetic, but when transformed into the beta form, the magnetism disappears although the lattice remains unchanged. The relations of these forms are peculiar. Pig iron is an alloy containing about 3 percent carbon with varying amounts of sulfur, silicon, manganese, and phosphorus [3].

Iron is a hard, brittle and can molded into many different forms. Iron is used to produce other alloys, including steel. It is the cheap, abundant, useful, and important metal. Wrought iron contains only a few tenths of a percent of carbon, is tough, malleable, less fusible, and usually has a "fibrous" structure. However, Carbon steel is an alloy of iron with small amounts of Mn, S, P, and Si. Alloy steels are carbon steels with other additives such as nickel, chromium, vanadium, etc.

2. Iron in cellular milieu (forms, states, and importance)

2.1 Iron forms

Cell is the structural and functional unit of life which performs many different physiological and biochemical processes that are essential for the survival of an organism, either unicellular or multicellular. The cells depend on food, essential micro and macro nutrients, vitamins, etc., to perform these biochemical processes in the tissues and organs of the body, and iron is one such essential requirement for cellular processes [4]. Living beings acquire the iron from the surrounding environment either by consumption (animals) or by absorption (prokaryotes, plants, fungi etc.,). The iron absorbed or taken through food exists in two important forms in the biological systems, the highly insoluble oxidized nonabsorbable form, $Fe3+$ and the readily absorbable reduced soluble form, $Fe2+$ [5]. These two forms also exist as free and bound forms and the levels of free and bound form is very critical for the normal functioning of the cells, opportunistic pathogen infections and the physiological state of cells or the iron related diseases of the organism [6–8]. For example, higher levels of free iron can help in bacterial infection and survival as in the case of Mycobacterium Tuberculosis [9] or the hemochromatosis (iron overload, a condition where body stores too much of iron which cause serious damage to the vital organs) [10]. The different optimum levels of iron in the body are illustrated in the **Figure 3**. So, the reduced $Fe2+$ form is the form that the cells can take up from the blood and is transported to many tissues where it is required [11]. The excess iron in the blood is absorbed by the cells and tissues is stored in the tissues or cells in a bound form, meaning that the iron is either incorporated into the enzymes, as cofactor, or can be stored in the tissues and cells in the body by special proteins called "Ferritins", or transported to the cells in bound form to iron transporters in the cells called "Transferrin" [12, 13].

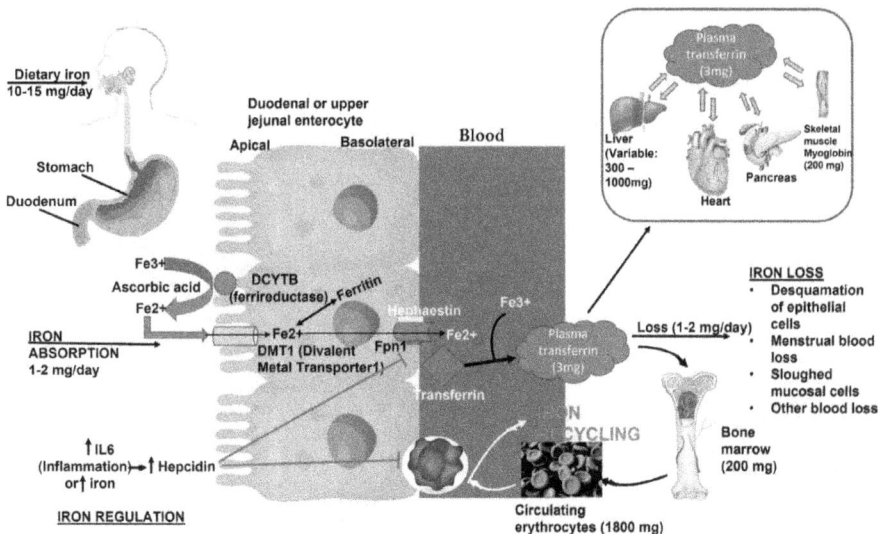

Figure 3.
Absorption of iron, transport, and the levels in the body. (ref. adapted from Jiten P Kothadia et al., 2016).

2.2 Levels and the distribution of iron in the body

The levels of iron in the blood, body tissue, and its distribution are very critical for the proper function, physiology, immune functions, and the determinant of the opportunistic infections in the body [14–16]. Hence, the levels of iron are critically regulated by the body tissues and cells. Here we shall investigate different levels and the distribution of iron, which control the tissue or the cellular absorption of the iron, by looking at the levels of iron in different tissues and the absorption of iron in human beings [17]. The normal levels of iron in the body are approximately 35 and 45 mg/kg body weight in adult women and men. Of the total iron we get form the food, about 60–70% is present in hemoglobulin in circulating RBCs, 10% is present in myoglobins (hemoglobin in the muscle), cytochromes, and iron containing enzymes accounting for not more than 4 mg–8 mg [18]. In healthy individual, 20–30% of iron is stored in the ferritins (an iron storage protein) and hemosiderins (iron acquisition protein) in hepatocytes and reticuloendothelial macrophages as shown in the **Figures 4** and **5**. Transferrin, another iron holding protein in the body, contains less than 1% (approx. 4 mg) of the total iron stores of the body and has the significant and highest turnover (25 mg/day). Transferrin transports 80% the iron in its bound form to majorly to the bone marrow for synthesis of hemoglobin in the development erythrocytes (Ibrahim Mustafa, 2011, Ph.D. thesis, http://hdl.handle.net/2429/36590) Thus, it transports the iron from blood to the bone marrow regularly and hence its high turnover. Most of the cellular iron is obtained from the dead RBCs and the iron released into the blood by reticuloendothelial cell macrophages in the liver where the senescence red blood cells are degraded after the completion of 120 days lifespan. So, the iron from the blood is regularly transported to all the tissue types or organelles by the transferrin molecules to deliver to the iron metalloprotein in those tissues. For example, the iron transported to mitochondria is incorporated into protoporphyrin IX, an important component of oxygen transport during oxidative phosphorylation. The iron which

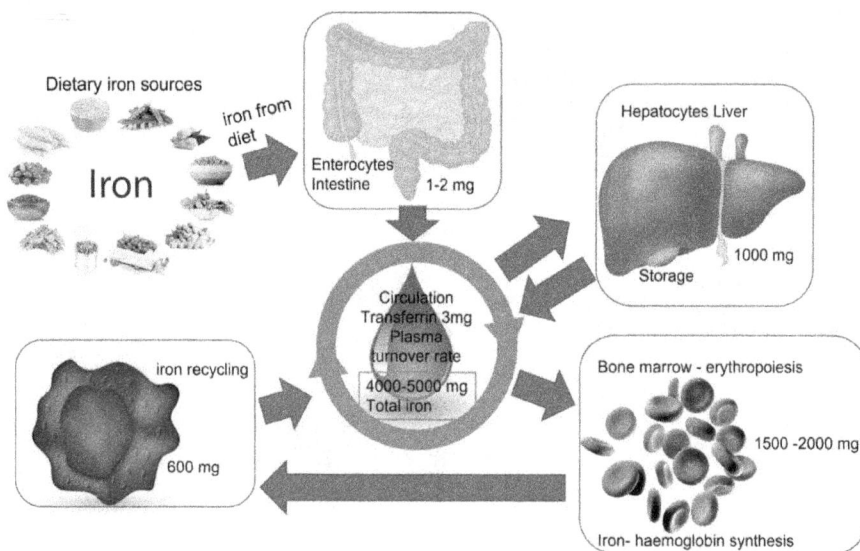

Figure 4.
Iron absorption and distribution in the body (adapted from Mustafa, 2011).

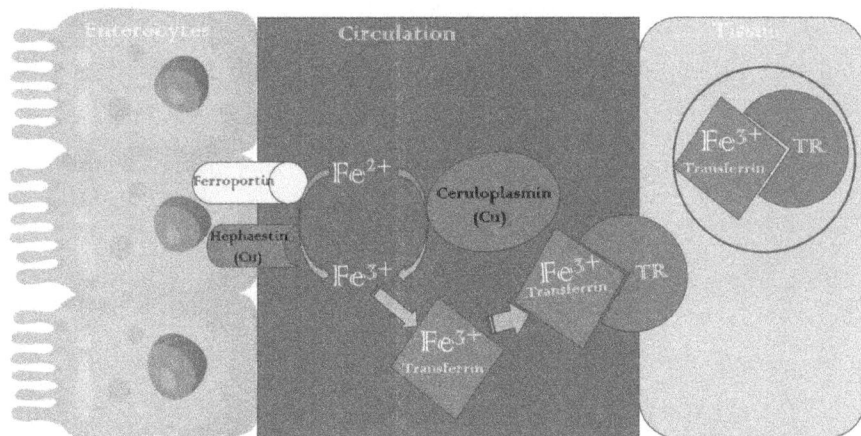

Figure 5.
Iron uptake by the cells or tissues and the regulation of iron uptake in the body (ref. Mustafa, 2011).

is lost due to wear and tear of the tissues and regularly during menstrual bleeding in females is replenished from the from dietary iron.

Iron is critical for oxygen transport and one of the most abundant metals in the human body, plays an important role in cellular processes such as the synthesis of DNA, RNA, and proteins that hep in the electron transport, cellular respiration, cell proliferation, differentiation, and regulation of gene expression [19]. Iron metabolism takes place in brain, testes, intestine, placenta, and skeletal muscles and high levels of iron is found in liver, brain, red blood cells, and macrophages [20]. Thus, the iron homeostasis is critical for the proper functioning of these organs and the altered levels results in the tissues results disturbed tissue functions and/or pathological or clinical conditions [11, 17]. Hence, it is very critical even during the blood transfusion to consider the levels of iron whether it is in bound or free form [21].

2.3 Iron transport in the cells

The dietary iron exists in two different forms, a haem form found in animal source foods which is the F2+ iron complexed with organic compounds and a non-haem form present in plant foods which is also called inorganic form. The non-haem iron or the Fe3+ iron is the major form of dietary iron, and this iron is not absorbed by the cells in the body, meaning that this Fe3+ needs to be reduced to Fe2+ to be taken up by the cells. This reduction is carried out the enzyme, ferric reductase also called ferric reductase duodenal cytochrome B (DCYTB or CYBRD1), which is found on the apical brush border membrane of intestinal epithelial cells in the duodenum and upper Jejunum [12]. The reduced Fe2+ is now transported or taken up by the apical surface of enterocytes lining the duodenal surface with the help of a 12 transmembrane divalent metal transporter protein1 (DMLT1). The Fe2+ iron transported to the enterocytes enters systemic circulation from the basolateral surface of the cells by the only known iron transporter, ferroportin another 12 transmembrane protein encoded by the Solute Carrier Family 40 Member 1(SLC40A1) gene. This transporter is also expressed in other cell types, particularly macrophages where it is highly expressed thus serves as iron uptake machinery which is utilized by the Mycobacterium Tuberculin during tuberculosis infection. Ferroportin also helps in the release of iron

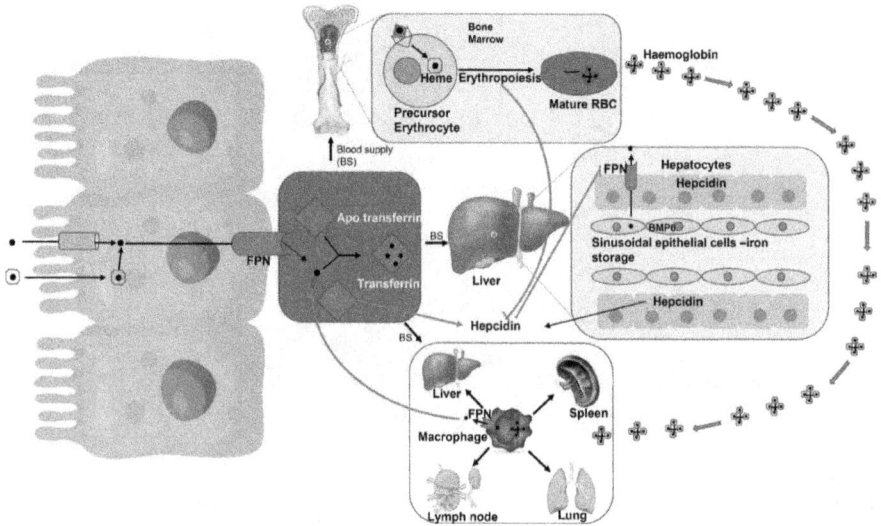

Figure 6.
Distribution of iron in the body for different function in the different tissues (picture taken from JBC MINIREVIEWS/ VOLUME 292, ISSUE 31, P12735–743).

form the stores from the hepatocytes assisted by the cupper containing ferroxidase enzyme ceruloplasmin or the membrane bound counterpart hephaestin, a membrane associated ferroxidase in the intestine (**Figure 6**). The enzymes oxidize the iron from Fe2+ form to Fe3+ form before it binds to transport protein transferrin which transports the iron to different tissues from the blood which will be discussing in the iron uptake by the cells or tissues [12].

3. Iron metabolism in cells (uptake, storage, transport, and component of cellular macromolecules)

Iron that is transported or absorbed into the system circulation is bound to the transporting protein, transferrin, present in the blood. Transferrin as the name suggests transfers the iron to all the tissues in the body. Iron also exists in the blood as non-transferrin bound form, especially when the serum levels iron is high, and the transferrin is completely saturated during hereditary haemochromatosis (HH) or any other iron overload conditions such as cancer, irregular heartbeat, and cirrhosis of the liver [22].

3.1 Uptake of iron by the cells

The transferrin bound iron in the blood is taken by the cells and tissues in the body. The transferrin along with its bound iron (holotransferrin) binds to the transferrin receptor1 (TFR1) that is expressed ubiquitously on the cell surfaces. The complex of iron-tranferrin-TFR1 is endocytosed by the cells. The endocytosed vesicles in the cell are acidified resulting in the opening of the DMT1 present in the vessels that releases the iron into the cells. The vesicle with the transferrin-TFR1 complex is recycled back to the membrane of the cells where it is reincorporated and thus the transferrin is released into the blood is available for the next round of iron transport basing on the levels iron in the blood. Iron is also transported to the cells through non-transferrin bound means as non-transferrin bound Iron (NTBI) which involves the zinc transporter ZRT/IRT-like protein 14 (ZIP14) a member of the SLC39A zinc transporter family. The iron that is transported to the cells and tissues is used for the cellular functions such as oxygen transport in RBCs by complexed with hemoglobin and myoglobin, cofactor for enzymes which are majorly involved in oxidation reduction reactions or stored as iron pools to serve the cell requirements when needed [5, 13].

3.2 Storage of iron in the tissues

Liver is the major site for iron storage in the body. The excess iron in the cells is stored in the iron storage proteins which contains iron holding pockets thousands in number to hold the iron. The iron storage proteins in the cells are ferritin and hemosiderin. Ferritin is the major iron storage protein at the cellular and organismal level. It stores 30% of the total storage iron in the cells and thus also sequesters the very reactive toxic fe2+ iron that generates reactive oxygen species (ROS) by Fenton reaction or subsequent reaction in the body. Ferritin is a spherical shell made up of 24 subunit proteins and has a centrally located iron holding cavity which can accommodate 4500 iron molecules in the Fe3 (III) complexed state. In this iron bound form ferritin not only stores the iron but also regulates the iron levels in the tissues by slow release (**Figure 7**). Hemosiderin is another protein which stores the iron in cells and tissues in the body. It is an iron storage complex of digested ferritin and lysosomes. Hemosiderin also forms when the ferritin is completely saturated with iron, the excess iron in the cells and tissues forms complex with phosphate and hydroxide forms.

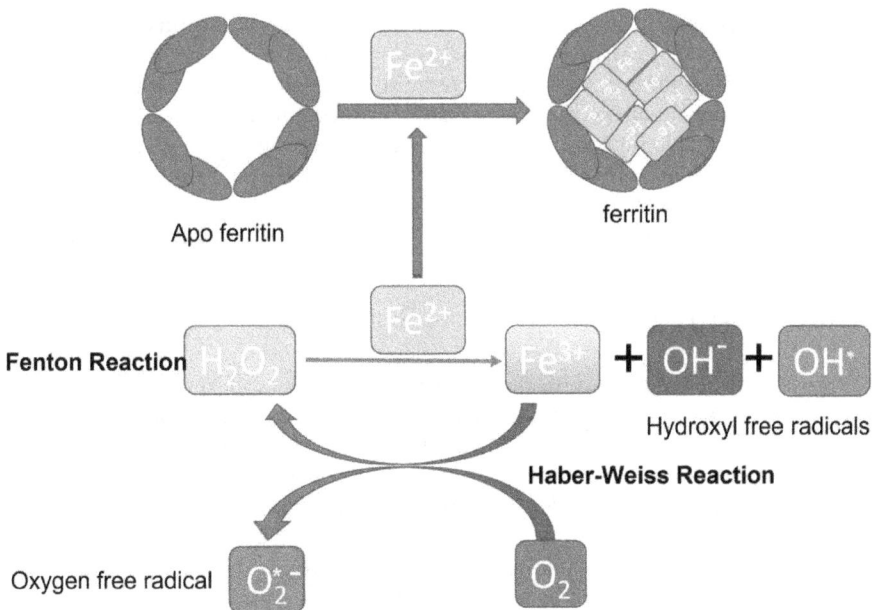

Figure 7.
Structure of ferritin and storage of iron. Fenton reaction to of generation of reactive of reactive oxygen species-ferritin sequesters excess iron and prevents in generating ROS. (ref: Antioxidants & Redox Signaling VOL. 10, NO. 6).

However, if the body burden of iron increases beyond normal levels, excess hemosiderin is deposited in the liver and heart. This can reach the point that the function of these organs is impaired, and death ensues [23].

3.3 Iron as a component of cellular macromolecules

About 95% of the total dietary iron in the cell of the body is in the bound form to proteins (about 95%) either in the cell cytosol or compartmentalized in different cell organelles. The rest 5% is available in form of free form represented as a cytosolic labile iron (LCI) or the labile pool iron [24]. This labile pool is, though very less, is the major for most of the deleterious effects in the body as it is very reactive, freely available, exchangeable and chelate able form in the cells. This labile form can initiate ROS generation, induce peroxidation of lipids of cells, or changes the oxidation and reaction in cells and thus is very toxic to cells. Here in this topic, we will discuss the important cellular components of iron and their physiological significance.

3.4 Labile cell iron (LCI)

LCI is the generic term for generic to describe labile iron in the cell, or cell compartments which exists in either fe2+ or Fe3+ form. The levels of LCI greatly vary depending on the location or cellular component, metabolic state, and the chemical composition of the component. The LCI is very important in the cell physiology as it serves as a metal source for metabolism also an indicator of cellular iron levels [25]. Thus, cells balance the uptake of circulating total bound iron (TBI) and store in the ferritin shells as unutilized iron basing on the LCI levels. Hence, this can be used as a dynamic cell parameter as the LCI pools are likely to vary over time in response to chemical or biological stimuli as well as to metabolic demands/responses. Conversely, the LCI levels decrease in iron starvation, which results in stable or transient over-expression of cytosolic or mitochondrial ferritin, a common scene in some mitochondrial disorders of aberrant mitochondrial iron accumulation. Thus, LCI acts an important indicator in pharmacological and research settings [24, 26]. LCI also helps in important components of chaperone role for human poly (rC)-binding proteins 1–4 (PCBPs 1–4), members of members of the heterogenous nuclear ribonucleoprotein family comprised of PCBPs 1–4 RNA/DNA-binding proteins involved in diverse processes such as splicing, transcript stabilization and translational regulation [27].

3.5 Mitochondrial iron metabolism

Mitochondrial plays and critical role in cellular iron metabolism as these are the only sites for heme synthesis, essential component of RBCs, skeletal or cardiac muscle oxygen carrying function and Iron–Sulfur cluster (ISC) biogenesis which serves a plethora of functions in the cells. These iron complexes are very critical component of metabolic enzymes, and defect is mitochondrial iron metabolism results in severe diseases [28]. Iron transport in mitochondria can be mediated by the kiss and run process of iron containing vesicles with mitochondrial membrane or the receptor mediated endocytosis which is mediated by particularly by PCBP2 [29]. Binding of this protein to the mitochondrial membrane DMT1 results in the efflux or the influx of iron into and out of mitochondria. Mitochondria also express its specific ferritin (FTMT) to store the iron [30]. High levels of FTMT are also expressed in sideroblasts (i.e., erythroblasts with iron granules) of patients affected by sideroblastic anemia [31].

3.6 Iron as component of heme

Heme is the precursor of oxygen carrying protein of the body's RBCs, skeletal and cardiac muscle, cytochromes, and many other enzymes. Heme contains 95% of functional iron in the human body, and two-thirds of the average person's dietary iron intake in developed countries and major cause of many of the iron associated diseases due to consumption of iron rich sources, especially from the animal origin [32]. It is complexed with porphyrin ring of hemoglobulin and myoglobin proteins. Apart from the enormous importance of iron or heme in the body, it is also important to note that the polymerization of heme (polyheme) which arises because of the neutralization of gastric contents by pancreatic juice or the Hemozoin that in formed during the erythrocytic cycle of *Plasmodium* infection in human beings are toxic to the cells (Modern Blood banking and transfusion practices, 6th edition) [33]. Besides, loss of iron because of tear and wear of tissues or the menstrual loss in human results in anemia. The list of diseases which are associated with the iron intake can be found at this reference (*Nutrients* 2014, 6, 1080–1102; doi:10.3390/nu6031080, [34]).

3.7 Iron as component of proteins

Eukaryotic cells contain numerous iron-containing proteins, which can be mainly classified into three groups: Iron–sulfur (Fe-S) cluster proteins, hemoproteins, and non-heme/non-Fe-S proteins. Fe-S proteins are characterized by their different structures with variable oxidation states, ranging from [2Fe-2S] diamonds, [3Fe-4S] intermediates, to [4Fe-4S] cluster cubes. Examples of Fe-S proteins include DNA polymerases, DNA helicases, hydrogenases, nicotinamide adenine dinucleotide (NADH)-dehydrogenases, nitrogenases, ferredoxins, and aconitases [35, 36]. Hemoproteins have a heme prosthetic group that allows them to carry out oxidative functions. Examples of hemoproteins include cytochromes, hemoglobin, myoglobin, catalases, and peroxidases. Nonheme/non-Fe-S proteins can be further subgrouped into three classes: Mononuclear non-heme iron enzymes, diiron proteins, and proteins involved in ferric iron transport. This group of iron-containing proteins mainly includes the small subunit of ribonucleotide reductases (RNRs), superoxide dismutases (SODs), dioxygenases, pterin-dependent hydrolases, and lipoxygenases. One can find the list of different proteins and their functions in cells at this reference (Zhang: Iron-containing proteins in Arabidopsis) [37]. Iron is also stored or absorbed by the microbiome in the gut plays a major role that not only influence the metabolism and genome of the host but also required for the growth of good bacteria and the harmful opportunistic bacteria that may cause dreadful disease.

4. Importance of iron in cellular processes (mechanism of homeostasis)

Iron being the essential micronutrient and a major component of cellular respiration, metabolism, its distribution in body fluids and tissue and the diseases associated with the alterations in the levels of free or bound iron results in many diseases ranging from anemia, hemochromatosis, and infections. It is obvious that the regulation of iron levels in the body is an important parameter of cellular state or understanding the physiology of the system, but here we shall discuss with some of clinical examples to iterate the point that the levels of iron in the tissues or the cells changes the cellular state or the diseases causing capacity of an organism that enters the host [27, 38]. We will not be discussing the iron homeostasis or regulation of iron uptake or absorption in the body.

4.1 Effect of iron levels in blood transfusions

I believe that each one of us are aware importance of blood transfusion in human life in the present days, which is essential in case of major surgeries, accidental blood loss, leukemia's, or the critical chronic blood infections. Blood group and Rh factor is the critical parameter during blood transfusions and the iron is no exception [39, 40]. It was found that the acute delivery of iron "predisposes patients to new infections, converts "benign" bacterial colonization into virulent infection, or enhances the virulence of existing infections" [41]. It should also be appreciated that the ease and the rapidity of the iron to interconvert between the ferric and ferrous forms interfere with the many of the cellular processes. The same excess iron in the blood also poses a potential threat to the cells as it initiates the production of reactive oxygen species (ROS) through Fenton reaction in the blood which would damage the cells or cause DNA damage in the delicate immune privileged organs, especially. Therefore, there is an active discussion on the levels of iron in the healthy blood donors, whether it is medically important and what steps to be taken while transfusions [42].

4.2 Effect of iron levels on the diseases causing pathogens

Iron availability to the parasites within the host is one of the critical factors that affects the disease-causing ability [43]. Mycobacterium tuberculosis is an intracellular pathogen that causes the life-threatening tuberculosis (TB) disease in host [44]. It has been found that the mutations that results in the macrophage iron overload results in the predisposition towards TB in HIV patients or increases the risk of TB.

It was also true for some of the commensal (microbiome), lactobacillus, Staphylococcus epidermidis or pathogens such as rabies can be deadly pathogens under right condition or the iron overload [45]. The reason is obvious as there is fierce competition between the host and the bacteria for iron under normal iron condition and this limits the iron for the bacteria to thrive to cause infection, but during the iron overload there is lot of iron available for these bacteria to convert into disease causing pathogens and innumerable pile up as we keep on the discussing this topic. Many more examples can be found in this editorial column (Iron: double edged sword).

	Men	Women
Functional iron		
Hemoglobin	32	28
Myoglobin	5	4
Fe-containing enzymes	1-2	1-2
Storage iron		
Ferritin and Hemosiderin	~ 11	~ 6
Transport iron		
Transferrin	0.04	0.04

Table 1.
Iron distribution in the adults (mg Fe/kg body weight) (Ref. Stipanuk MH (2006).

4.2.1 Effect of iron levels on immune system

Immune system in the body sequesters the iron in the body and thus prevents the iron dependent disease-causing pathogens, this is called the nutritional immunity [46]. In addition, nutritional immunity also modulates adaptive immune responses either in deficient or overload states [47].

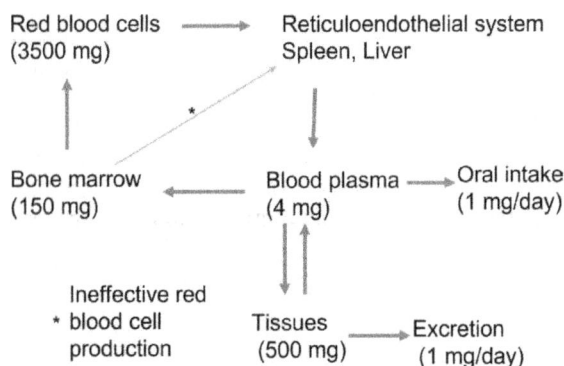

Figure 8.
Iron compartmentalization in the body (http://en.wikipedia.org/wiki/File:lron_metabolism.svg).

Confounder	Indicator and direction of change	Comment
Inflammation	SF ↑	Ferritin is a positive acute-phase protein
	Transferrin ↓	Transferrin is a negative acute-phase protein
	Iron ↓	The release of cytokines leads to increased uptake and retention of iron in reticuloendothelial system cells, e.g., iron becomes sequestered and is not available for transport to the bone marrow for erythropoiesis
	EP ↑	
	Hemoglobin ↓	
Increased erythropoietic activity	EP, sTfR ↑	In thalassemia, sickle cell anemia, and hemoglobinopathies
Lead poisoning	EP ↑	Lead blocks the formation of heme and zinc protoporphyrin forms instead
Pregnancy	Hemoglobin ↓	Plasma volume expansion results in hemodilution
Dehydration	Hemoglobin ↑	The volume of fluid in blood drops and hemoglobin artificially rises
Smoking	Hemoglobin ↑	Compensation for decreased oxygen intake in heavy smokers
Altitude	Hemoglobin ↑	Compensation for decreased oxygen intake due at high altitude

EP, erythrocyte protoporphyrin; SF, serum ferritin; sTfR, soluble transferrin receptor;, increase in concentration;, decrease in concentration.

Table 2.
Important confounders of iron status indicators (Ref. Am J Clin Nutr 2017;106(Suppl):1606S–14S).

4.2.2 Measuring or quantification methods of iron

Iron being the important element in different cellular processes, components of cells, enzymes, and proteins and the levels of iron are very well regulated in the body. The changes in the local concentrations in the cells and tissues results in many physiological and pathological conditions. Hence, estimating the concentrations acts as a good indicator to understand cellular physiological status and the pathological conditions in the body [48].

There are three main iron compartments in the and the alterations in the normal in each of the compartment is used as an indicator and the biochemical assessment is based on the iron levels in the serum. Thus offers an easy way to assess the iron concentrations in the serum that gives the readout about the infection state or the tissue physiology. Below are the irons indicators and the different understanding of the different physiological state of the body or tissues with respect to the iron alterations in the specific compartment (**Table 1** and **Figure 8**) and the condition (**Table 2**).

5. Conclusion

Iron is the essential micronutrient for the cell physiology and function that also points out not just the role in cellular function, but also a critical component in cellular infection, cytotoxicity, and the generation of reactive oxygen levels. This chapter provides the basic importance about the cellular iron, absorption, distribution, storage, critical concentration of tissue iron in the body and hints about the importance of these critical levels in the diseases and pathologies. Iron as an essential nutrient in the body and the increased levels or the free in iron in the body are more damaging to the cell and hence the iron is aptly called as double-edged sword. This chapter has discussed the pathogenesis of some normal gut microbes turning into pathogenic sps, macrophage iron levels and the infection of mycobacterium, haemochromatosis are the few examples that indicate the critical levels of iron in cell physiology and function. The critical importance of iron and this chapter provides the readers the importance of iron levels and points out to the fact that the iron has lot of scope in terms of understanding cell physiology, defining the cell function in diseases. Thus, the iron offers a huge scope for the research towards limiting the survival of pathogens in the body or could enhance the survival of good bacteria in the gut.

Besides, consideration regarding the blood transfusion, the irons levels (bound and unbound form in the donor blood would a major factor in blood transfusion, especially in terms of treating anemic patients, personalized medicine as in the cases of bacterial infection as to many species have differential sensitivity or pathogenicity to the iron levels are the areas of active debate and research.

Iron in Cell Metabolism and Disease
DOI: http://dx.doi.org/10.5772/intechopen.101908

Author details

Eeka Prabhakar
GITAM (Deemed to be University), Visakhapatnam, AP, India

*Address all correspondence to: eekabiotek@gmail.com

IntechOpen

References

[1] Youssef LA, Spitalnik SL. Iron: A double-edged sword. Transfusion. 2017;**57**(10):2293-2297

[2] Stixrude L, Cohen RE. High-pressure elasticity of iron and anisotropy of earth's inner core. Science. 1995;**267**(5206):1972-1975

[3] Avner SH. Introduction to Physical Metallurgy. New York: McGraw-Hill; 1964. pp. vii, 536

[4] Sigel A, Sigel H. Metal ions in biological systems, volume 35: Iron transport and storage microorganisms, plants, and animals. Met Based Drugs. 1998;**5**(5):262

[5] Lane DJ, Merlot AM, Huang ML, Bae DH, Jansson PJ, Sahni S, et al. Cellular iron uptake, trafficking and metabolism: Key molecules and mechanisms and their roles in disease. Biochimica et Biophysica Acta. 2015;**1853**(5):1130-1144

[6] Johnstone D, Milward EA. Molecular genetic approaches to understanding the roles and regulation of iron in brain health and disease. Journal of Neurochemistry. 2010;**113**(6): 1387-1402

[7] Goswami T, Rolfs A, Hediger MA. Iron transport: Emerging roles in health and disease. Biochemistry and Cell Biology. 2002;**80**(5):679-689

[8] Lieu PT, Heiskala M, Peterson PA, Yang Y. The roles of iron in health and disease. Molecular Aspects of Medicine. 2001;**22**(1-2):1-87

[9] Pesek J, Buchler R, Albrecht R, Boland W, Zeth K. Structure and mechanism of iron translocation by a Dps protein from Microbacterium arborescens. The Journal of Biological Chemistry. 2011;**286**(40):34872-34882

[10] Khan FA, Fisher MA, Khakoo RA. Association of hemochromatosis with infectious diseases: Expanding spectrum. International Journal of Infectious Diseases. 2007;**11**(6):482-487

[11] MacKenzie EL, Iwasaki K, Tsuji Y. Intracellular iron transport and storage: From molecular mechanisms to health implications. Antioxidants & Redox Signaling. 2008;**10**(6):997-1030

[12] Knutson MD. Iron transport proteins: Gateways of cellular and systemic iron homeostasis. The Journal of Biological Chemistry. 2017;**292**(31): 12735-12743

[13] Waldvogel-Abramowski S, Waeber G, Gassner C, Buser A, Frey BM, Favrat B, et al. Physiology of iron metabolism. Transfusion Medicine and Hemotherapy. 2014;**41**(3):213-221

[14] Soares MP, Weiss G. The iron age of host-microbe interactions. EMBO Reports. 2015;**16**(11):1482-1500

[15] Weinberg ED. Nutritional immunity. Host's attempt to withold iron from microbial invaders. Journal of the American Medical Association. 1975;**231**(1):39-41

[16] Soares MP, Hamza I. Macrophages and iron metabolism. Immunity. 2016;**44**(3):492-504

[17] Kothadia JP, Arju R, Kaminski M, Mahmud A, Chow J, Giashuddin S. Gastric siderosis: An under-recognized and rare clinical entity. SAGE Open Medicine. 2016;**4**:2050312116632109

[18] Hamilton JL, Kizhakkedathu JN. Polymeric nanocarriers for the treatment of systemic iron overload. Molecular and Cell Therapy. 2015;**3**:3

[19] Boldt DH. New perspectives on iron: An introduction. The American Journal of the Medical Sciences. 1999;**318**(4):207-212

[20] Tandara L, Salamunic I. Iron metabolism: Current facts and future directions. Biochemia Medica (Zagreb). 2012;**22**(3):311-328

[21] Andrews NC. Forging a field: the golden age of iron biology. Blood. 2008;**112**(2):219-230

[22] Shawki A, Knight PB, Maliken BD, Niespodzany EJ, Mackenzie B. H(+)-coupled divalent metal-ion transporter-1: Functional properties, physiological roles and therapeutics. Current Topics in Membranes. 2012;**70**:169-214

[23] Wallace DF. The regulation of iron absorption and homeostasis. Clinical Biochemist Reviews. 2016;**37**(2):51-62

[24] Cabantchik ZI. Labile iron in cells and body fluids: Physiology, pathology, and pharmacology. Frontiers in Pharmacology. 2014;**5**:45

[25] Kakhlon O, Cabantchik ZI. The labile iron pool: Characterization, measurement, and participation in cellular processes. Free Radical Biology & Medicine. 2002;**33**(8):1037-1046

[26] Kruszewski M. Labile iron pool: The main determinant of cellular response to oxidative stress. Mutation Research. 2003;**531**(1-2):81-92

[27] Lan P, Pan KH, Wang SJ, Shi QC, Yu YX, Fu Y, et al. High serum iron level is associated with increased mortality in patients with sepsis. Scientific Reports. 2018;**8**(1):11072

[28] Horowitz MP, Greenamyre JT. Mitochondrial iron metabolism and its role in neurodegeneration. Journal of Alzheimer's Disease. 2010;**20**(Suppl. 2): S551-S568

[29] Chen C, Paw BH. Cellular and mitochondrial iron homeostasis in vertebrates. Biochimica et Biophysica Acta. 2012;**1823**(9):1459-1467

[30] Richardson DR, Lane DJ, Becker EM, Huang ML, Whitnall M, Suryo Rahmanto Y, et al. Mitochondrial iron trafficking and the integration of iron metabolism between the mitochondrion and cytosol. Proceedings of the National Academy of Sciences of the United States of America. 2010;**107**(24):10775-10782

[31] Sheftel AD, Richardson DR, Prchal J, Ponka P. Mitochondrial iron metabolism and sideroblastic anemia. Acta Haematologica. 2009;**122**(2-3): 120-133

[32] Abbaspour N, Hurrell R, Kelishadi R. Review on iron and its importance for human health. Journal of Research in Medical Sciences. 2014;**19**(2):164-174

[33] Olafson KN, Ketchum MA, Rimer JD, Vekilov PG. Mechanisms of hematin crystallization and inhibition by the antimalarial drug chloroquine. Proceedings of the National Academy of Sciences of the United States of America. 2015;**112**(16):4946-4951

[34] Hooda J, Shah A, Zhang L. Heme, an essential nutrient from dietary proteins, critically impacts diverse physiological and pathological processes. Nutrients. 2014;**6**(3):1080-1102

[35] Liu J, Chakraborty S, Hosseinzadeh P, Yu Y, Tian S, Petrik I, et al. Metalloproteins containing cytochrome, iron-sulfur, or copper redox centers. Chemical Reviews. 2014;**114**(8):4366-4469

[36] Andreini C, Putignano V, Rosato A, Banci L. The human iron-proteome. Metallomics. 2018;**10**(9):1223-1231

[37] Zhang C. Involvement of iron-containing proteins in genome integrity in arabidopsis thaliana. Genome Integrity. 2015;**6**:2

[38] Li Y, Huang X, Wang J, Huang R, Wan D. Regulation of iron homeostasis and related diseases. Mediators of Inflammation. 2020;**2020**:6062094

[39] Hod EA, Zhang N, Sokol SA, Wojczyk BS, Francis RO, Ansaldi D, et al. Transfusion of red blood cells after prolonged storage produces harmful effects that are mediated by iron and inflammation. Blood. 2010;**115**(21): 4284-4292

[40] Cable RG, Glynn SA, Kiss JE, Mast AE, Steele WR, Murphy EL, et al. Iron deficiency in blood donors: The REDS-II Donor Iron Status Evaluation (RISE) study. Transfusion. 2012;**52**(4): 702-711

[41] Natanson C, Danner RL, Elin RJ, Hosseini JM, Peart KW, Banks SM, et al. Role of endotoxemia in cardiovascular dysfunction and mortality. Escherichia coli and Staphylococcus aureus challenges in a canine model of human septic shock. The Journal of Clinical Investigation. 1989;**83**(1):243-251

[42] Solomon SB, Wang D, Sun J, Kanias T, Feng J, Helms CC, et al. Mortality increases after massive exchange transfusion with older stored blood in canines with experimental pneumonia. Blood. 2013;**121**(9):1663-1672

[43] Wang SC, Lin KH, Chern JP, Lu MY, Jou ST, Lin DT, et al. Severe bacterial infection in transfusion-dependent patients with thalassemia major. Clinical Infectious Diseases. 2003;**37**(7):984-988

[44] Kolloli A, Singh P, Rodriguez GM, Subbian S. Effect of iron supplementation on the outcome of non-progressive pulmonary mycobacterium tuberculosis infection. Journal of Clinical Medicine. 2019;**8**(8):1155

[45] Nairz M, Schroll A, Haschka D, Dichtl S, Tymoszuk P, Demetz E, et al. Genetic and dietary iron overload differentially affect the course of salmonella typhimurium infection. Frontiers in Cellular and Infection Microbiology. 2017;**7**:110

[46] Iatsenko I, Marra A, Boquete JP, Pena J, Lemaitre B. Iron sequestration by transferrin 1 mediates nutritional immunity in Drosophila melanogaster. Proceedings of the National Academy of Sciences of the United States of America. 2020;**117**(13):7317-7325

[47] Cherayil BJ. Iron and immunity: Immunological consequences of iron deficiency and overload. Archivum Immunologiae et Therapiae Experimentalis (Warsz). 2010;**58**(6): 407-415

[48] Pfeiffer CM, Looker AC. Laboratory methodologies for indicators of iron status: Strengths, limitations, and analytical challenges. The American Journal of Clinical Nutrition. 2017;**106**(Suppl. 6):1606S-1614S

Chapter 3

Role of Transferrin in Iron Metabolism

Nitai Charan Giri

Abstract

Transferrin plays a vital role in iron metabolism. Transferrin is a glycoprotein and has a molecular weight of ~80 kDa. It contains two homologous iron-binding domains, each of which binds one Fe (III). Transferrin delivers the iron to various cells after binding to the transferrin receptor on the cell surface. The transferrin-transferrin receptor complex is then transported into the cell by receptor-mediated endocytosis. The iron is released from transferrin at low pH (e.g., endosomal pH). The transferrin-transferrin receptor complex will then be transported back to the cell surface, ready for another round of Fe uptake and release. Thus, transferrin plays a vital role in iron homeostasis and in iron-related diseases such as anemia. In the case of anemia, an increased level of plasma transferrin is often observed. On the other hand, low plasma transferrin level or transferrin malfunction is observed during the iron overdose. This chapter will focus on the role of transferrin in iron metabolism and diseases related to transferrin.

Keywords: Transferrin, metabolism, transferrin receptor, homeostasis, endocytosis, intestine, divalent metal transporter (DMT1), Steap3, endosome

1. Introduction

Iron metabolism is one of the most intricate processes involving many organs and tissues, such as the intestine, the bone marrow, the spleen, the liver, etc. [1, 2]. Various proteins are also involved in maintaining iron homeostasis. Transferrin is a glycoprotein that plays a central role in iron metabolism [3]. It is present at a concentration of 30-60 μM in blood [4]. Transferrin can be divided into several sub-groups – serum transferrin, lactoferrin, and ovotransferrin. Hepatocytes produce serum transferrin found in serum, CSF, etc. Mucosal epithelial cells produce lactoferrin found in milk [5]. Lactoferrin is also found in secretions such as tear and saliva and cells such as neutrophils and leukocytes. Ovotransferrin is an iron-binding protein found in avian egg white. Together transferrins form the most important iron regulation system by transporting iron from the intestine or the sites of heme degradation to proliferating cells [6, 7]. This chapter will focus on the role of transferrin in iron metabolism.

Unlike ferritin, transferrin is a relatively new protein and is found only in phylum Chordata. Transferrin contains ~680 amino acid residues and two subdomain (N-and C-terminal domains) or lobes (**Figure 1**) [8]. It has a molecular weight of 80 kDa.

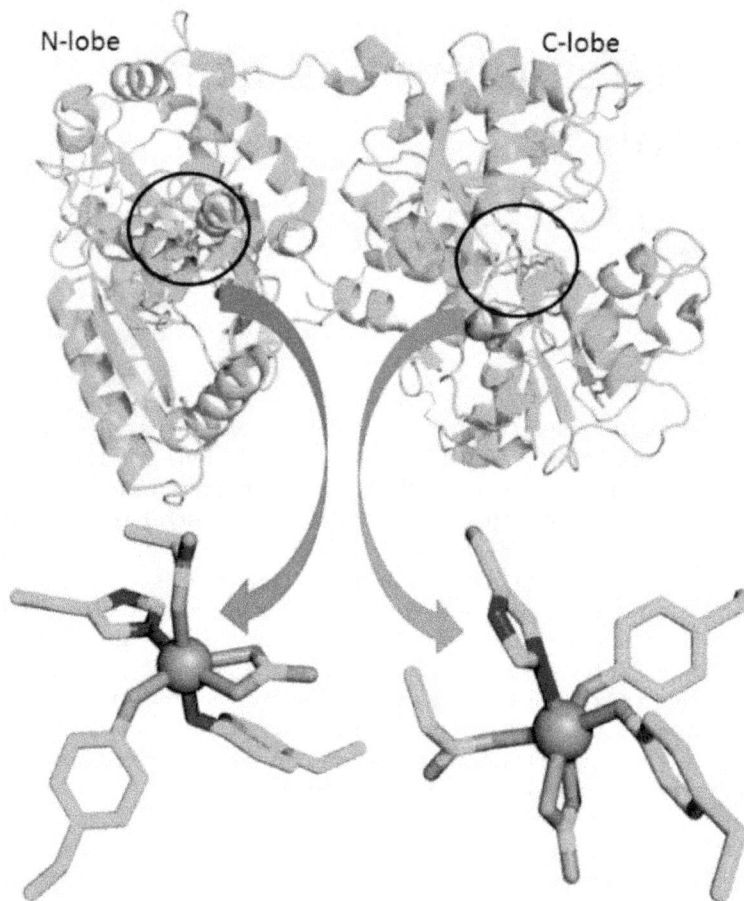

Figure 1.
Structure of human transferrin (top) and two of its iron (brown sphere)-binding sites (bottom).

The N-terminal lobe consists of residues 1-330 (approx.), while the C-terminal lobe consists of residues 340-680 (approx.) [9]. The two subunits are connected by a small hinge (residues 330-340). Transferrins show high sequence similarity - ~70% identity among lactoferrins while 50-60% identity between lactoferrin and transferrin [10]. The N- and C-terminal halves of these molecules show ~40% sequence identity. It has been suggested that the transferrin molecule may have evolved from the structural gene of an ancestral protein possessing only one metal-binding site and about 340 amino acids by gene duplication [11, 12]. This gene duplication might have led to an increase in its Fe(III) binding capacity and affinity [13]. Although transferrin contains many cysteine residues, it does not have any free sulfhydryl groups (present as disulfide).

Transferrins show greater species variability in carbohydrate composition than in their amino acid composition. The total carbohydrate content varies from 3–12% weight of protein. The number of carbohydrate chains per protein molecule varies from 1 to 4. Human serum transferrin contains about 6% carbohydrate. This carbohydrate moiety has two identical, branched hetero-saccharide chains attached to the amide group of Asn residues via N-glycosidic linkages. However, a minor population of transferrin contains only tri-branched glycans. These carbohydrate groups affect the recognition and conformation of the native protein. The carbohydrate groups can also influence the solubility of the protein.

2. Iron binding to transferrin

Transferrins are bilobal where each lobe reversibly binds a ferric iron (logK = 22.5 for C-site and 21.4 for N-site). Although two iron sites can be distinguished by kinetic and few other studies, their coordination environments are similar (**Figure 1**, bottom). X-ray crystallography indicates that the iron-binding site involves two phenolate oxygen from Tyr, two oxygen from bidentate bicarbonate, nitrogen from His, and oxygen from the carboxylate group of an Asp. Although transferrin binds Fe(III), iron is absorbed as Fe(II) in the intestine. Ceruloplasmin may catalyze the oxidation of Fe(II) to Fe(III) so that it may be bound to transferrin. However, there is some evidence that transferrin binds Fe(II), although with a much lower affinity. The resulting Fe(II)-transferrin-bicarbonate (or carbonate) will be oxidized by molecular oxygen to Fe(III)-transferrin-bicarbonate (or carbonate) [14, 15]. Thus, transferrin binds two Fe(III) in the presence of carbonate or bicarbonate to form a pink-colored complex with an absorption maximum of 465-470 nm. This iron-binding is pH-dependent, where the efficiency of iron-binding is maximum at pH between 7.5 and 10. This iron-binding efficiency decreases upon lowering the pH, and partial dissociation occurs at pH 6.5. Complete dissociation of iron occurs at pH 4.5. This decrease in iron-binding efficiency is useful for preparing apo-transferrin *in vitro*. For every Fe(III) bound to

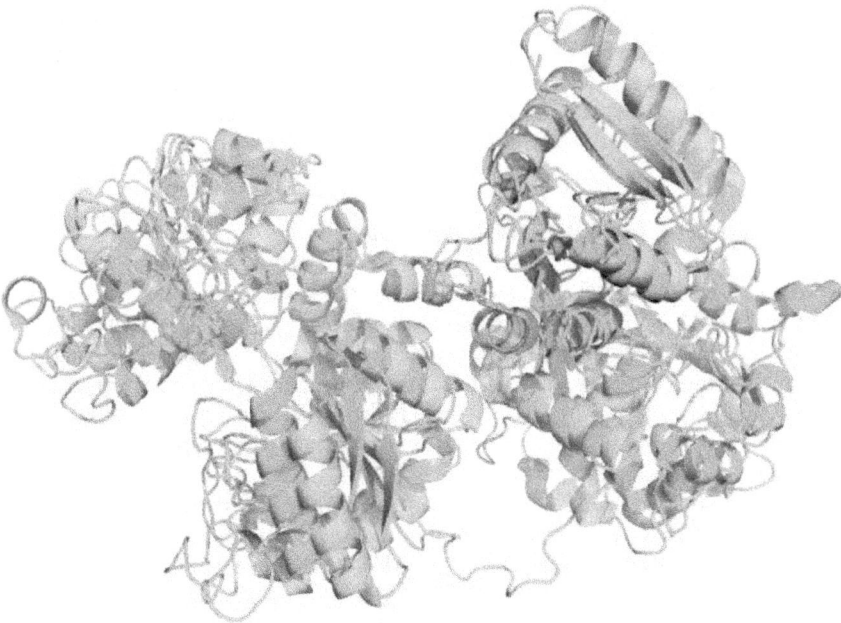

Figure 2.
Superimposition of the structures of apo-transferrin (cyan) and holo(diferric)-transferrin (green).

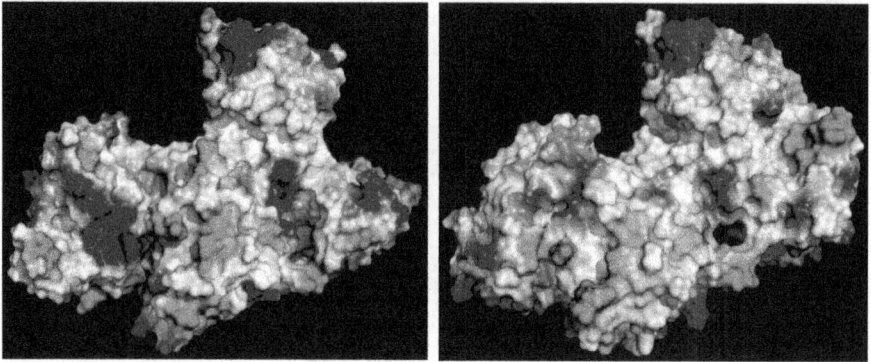

Figure 3.
Distribution of surface charge in apo-transferrin (left) and holo-transferrin (right).

the protein, three protons are released. Considering that two Tyr residues are bound to Fe(III), these two ligands may be responsible for two protons. The third proton may come from bicarbonate.

Superimposition of the apo-transferrin (no iron) structure with the holo form (diferric) indicates the presence of open and closed forms, respectively (**Figure 2**). It has been suggested that apo-transferrin can exist both in an open and closed formation. However, the closed conformation exists less than 10% of the time [16]. Differential scanning calorimetry experiment performed by titrating Fe(III) into apo-transferrin indicated cooperativity between the two lobes in transferrin. During this process, Fe(III) first binds to the C-lobe and then to the N-lobe. It was also observed that the binding of Fe(III) in the C-lobe helps strengthen the binding of Fe(III) in N- lobe. It is worth noting that the interface of the lobes contains hydrophobic patches. The hydrophobic interaction may cause the movement in one lobe as the other one closes due to Fe(III) binding. Also, Fe(III) binding to transferrin alters the surface charge (**Figure 3**). The surface charge in holo-transferrin is more negative than apo-transferrin. Thus, electrostatics may drive the onset of endocytosis.

3. Transfer of iron from transferrin to cells

Under physiological conditions, Fe(III) is tightly bound to transferrin. Considering the K_a of the reaction between transferrin and iron, iron will take thousands of years to dissociate spontaneously from transferrin to blood. Thus, there must be some unique mechanism for transferring iron from transferrin to cells. Now it is established that transferrin receptors on the cell surface play an important role in transferring iron from transferrin to cells. Transferrin receptor is a 180 kDa homodimer type II transmembrane glycoprotein [17]. Two monomers of ~769 amino acids are linked by two disulfide bridges [18]. Each monomer contains three major domains – a C-terminal extracellular domain, a transmembrane domain, and an N-terminal intracellular domain. The C-terminal extracellular domain comprises ~671 amino acids and has two main subunits – domain head and a stock of ~37 amino acids that separate the head from the transmembrane domain. The transmembrane domain consists of ~20-28 residues that create a hydrophobic region. This region contains palmitoylation sites (Cys62 and Cys67) that help the transferrin receptor cling to the cell membrane. The intracellular N-terminal domain is made up of ~61-66 residues. Due to the dimeric nature of the transferrin receptor, it binds two transferrin molecules.

3.1 Conformational change in the transferrin receptor due to transferrin binding

X-ray crystal structure of iron-bound transferrin in complex with transferrin receptor provides the molecular details of the interaction between transferrin and transferrin receptor (**Figure 4**) [19]. Conformation changes in the C-terminal domain of transferrin receptor occurs when transferrin receptor binds iron-bound transferrin. These conformation changes have been suggested to be responsible for initiating endocytosis for Fe(III) uptake by the cells. The most dramatic change in the transferrin receptor structure is observed in the loop containing Asn317 (one of the three glycosylation sites, **Figure 5A**). Also, Phe316 is shifted by ~8 Å while His318 is shifted by ~12.5 Å. These movements bring these residues closer to the C-terminus of other transferrin receptor monomer (**Figure 5B**). The binding of transferrin to transferrin receptor leads to a rotation along transferrin receptor dimeric interface bringing four His (His475 and His684 from each monomer) into proximity (**Figure 5C**). The binding of transferrin to transferrin receptor also causes two Trp residues (Trp641 and Trp740) to undergo significant changes (**Figure 5D**).

The interaction of the transferrin receptor with transferrin also depends on pH. For these interactions, the important pH values are 7.4 (blood pH) and 5.5 (endosomal pH). It has been reported that at blood pH diiron bound transferrin exclusively formed saturated transferrin-transferrin receptor complex [20]. However, monomeric

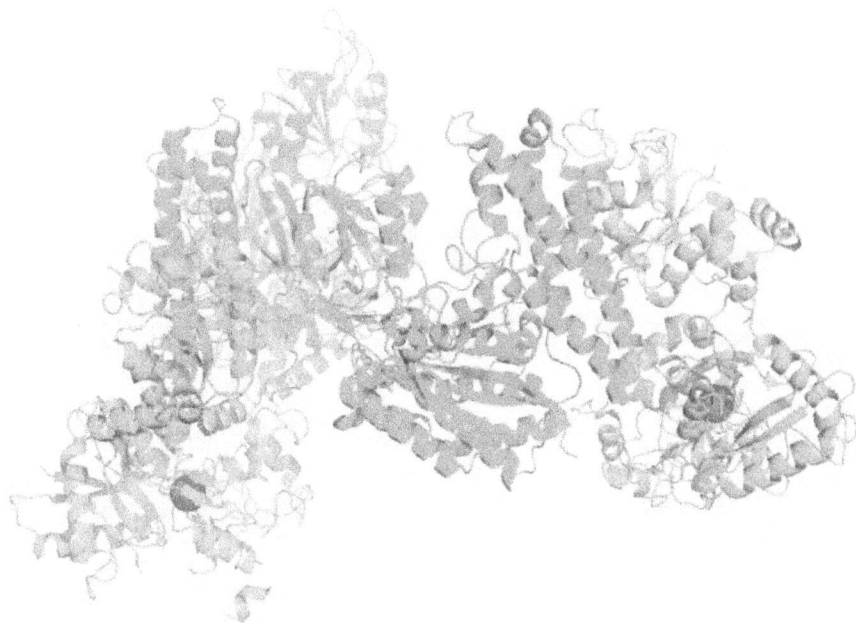

Figure 4.
Interaction of transferrin (cyan: The brown sphere is transferrin-bound iron) with C-terminal extracellular domain of transferrin receptor (green).

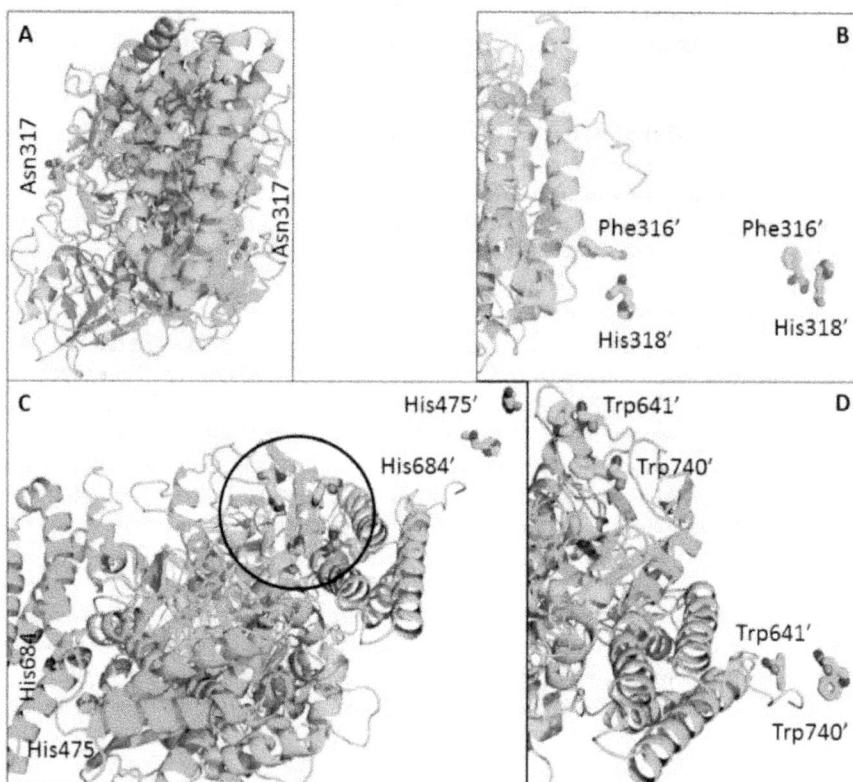

Figure 5.
Conformational change of transferrin receptor (cyan) due to the formation of transferrin-transferrin receptor complex (green): Structural change of the loop containing Asn317 (A), movements of Phe316 and His318 from one monomer (indicated by ') to the other monomer (B) four his residues of transferrin receptor come to close proximity due to transferrin binding (C) and movements of Trp641 and Trp740 from one monomer (indicated by ') to the other monomer (D).

transferrin (N-lobe or C-lobe) with one iron as well as apo-transferrin could not saturate the transferrin receptor. It was shown that the diiron-containing transferrin has the maximum affinity for the transferrin receptor, while the apo-transferrin had the lowest affinity for the transferrin receptor. However, this trend is reversed at endosomal pH. Here, apo-transferrin has the highest affinity for the transferrin receptor. This strong interaction is necessary for transferrin to return to the cell surface and repeat further Fe(III) uptake.

Thus, the recognition of diiron bound transferrin is essential for initiating endocytosis (**Figure 6**). This transferrin-mediated endocytosis involves the clathrin coating of the ensuing endosome to protect it from proteolytic degradation. Thus, the enclosed transferrin and transferrin receptor are protected so that they can be recycled. The adaptor protein complex mediates the formation of a proton-pumping endosome that includes other membrane proteins such as Steap3, a ferrireductase [21, 22]. Once the endosome enters the cell, it is acidified to a pH of 5.5. This acidification may enable the chelator to penetrate the metal-binding site and induce a semi-open conformation that ultimately leads to metal release. It has been reported that the sulfate binding to Fe(III) prevents His and Asp from binding to Fe(III) and thus leads to a semi-open conformation of the transferrin (**Figure** 7). During the metal release and its delivery to the cytosol via the divalent metal transporter 1 (DMT1), the reduction of Fe(III) to Fe(II) occurs. However, there is some debate about the order of chelation

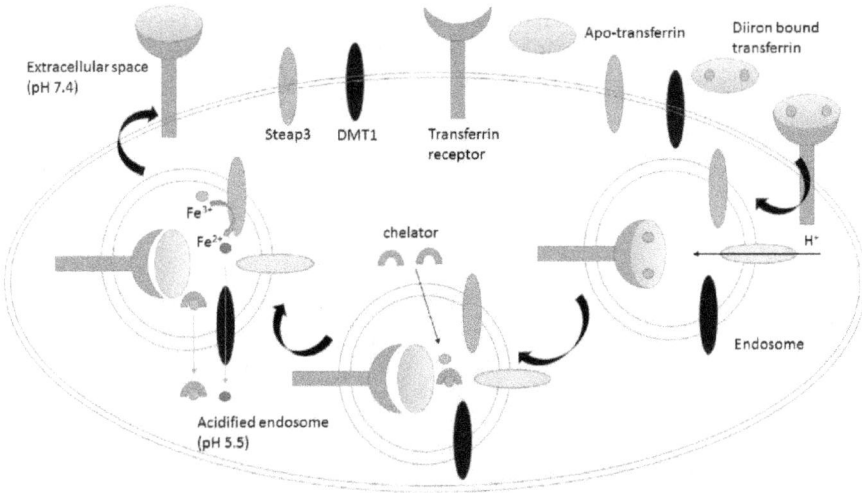

Figure 6.
Proposed pathway of transferrin receptor mediated endocytosis and iron release into the cytosol.

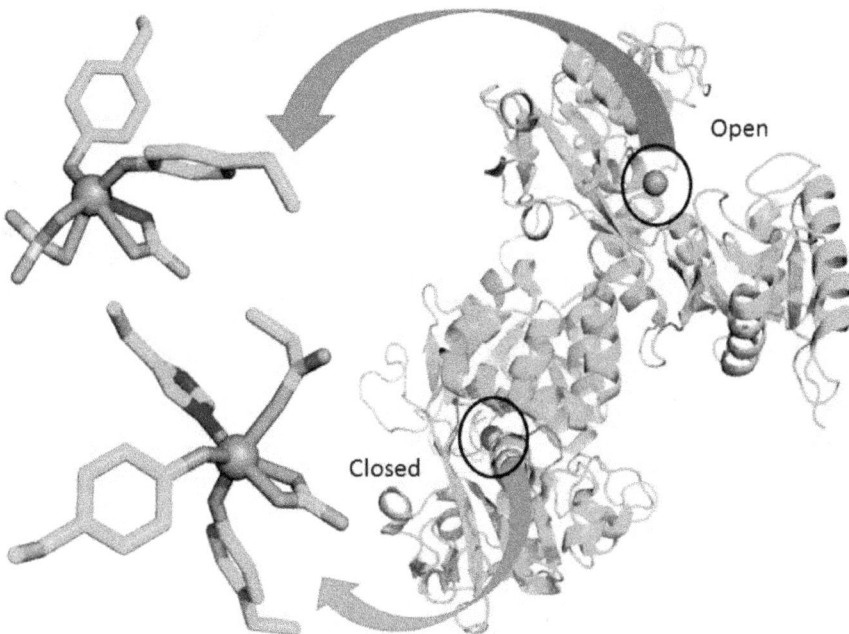

Figure 7.
Sulfate binding to iron (brown sphere) leads to semi-open conformation by preventing the ligation of his and asp to iron.

and reduction. One group of researchers think that acidification coupled with chelation by an intracellular chelator (e.g., citrate, ATP) results in the dissociation of Fe(III) from transferrin [23]. This Fe(III) is then reduced to Fe(II) by Steap3. DMT1 then transports Fe(II) into the cytosol, forming a labile iron pool [24, 25]. However, this Fe(II) is quickly stored in ferritin and inserted into various iron-dependent proteins [26, 27]. Another group of researchers believes that the interaction between diiron bound transferrin and transferrin receptor alters the redox potential of Fe(III) from −0.53 V to −0.3 V (vs. SHE) [28, 29]. Then Steap3 will reduce Fe(III) to Fe(II), which

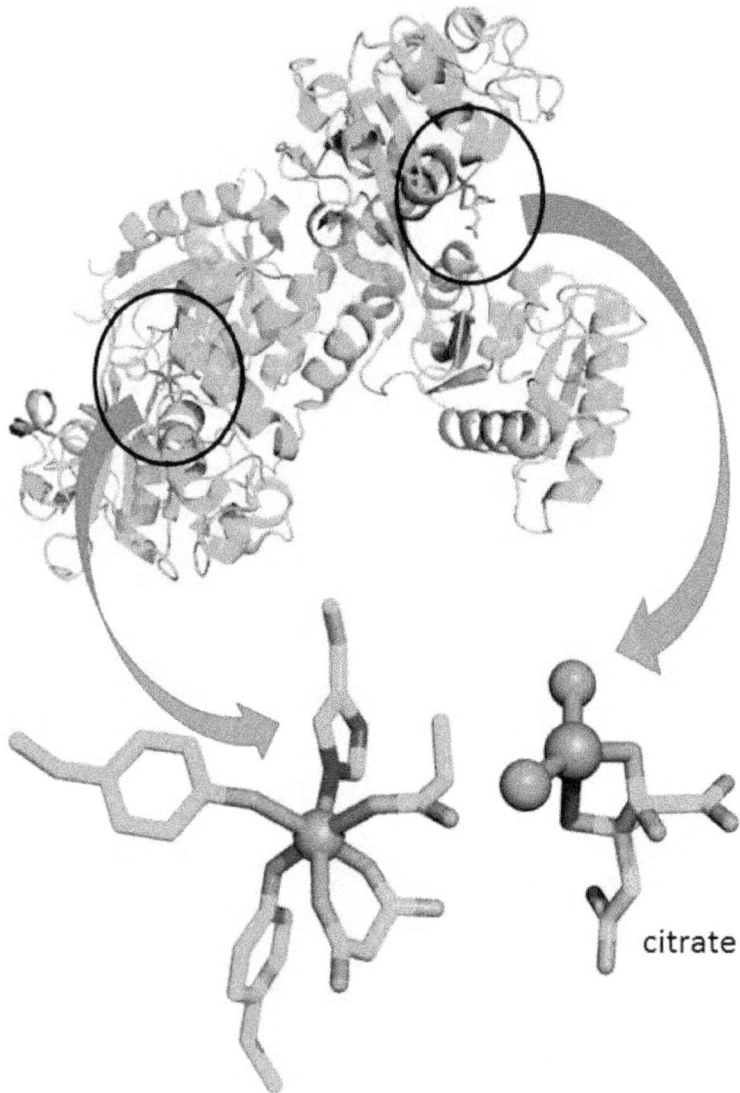

citrate

Figure 8.
Structure of citrate bound iron (brown sphere) near the metal binding site in transferrin.

will weaken the metal affinity of transferrin (logβ value of diFe(III) bound transferrin is 43.5 while that of diFe(II) bound transferrin is 13) [30, 31]. Fe(II) will then undergo facile dissociation, possibly with the help of a chelator, and be transported out of the endosome by DMT1. Some researchers also believe that ascorbate is the likely reducing agent [32]. Recently, Fe(III) bound to citrate near the transferrin metal-binding site has been reported (**Figure 8**). However, in this structure, Fe(III) is not ligated to any protein-derived ligand. This citrate-bound iron (without protein-derived ligand) can be considered citrate scavenging of Fe(III) from transferrin.

4. Significance of bilobal transferrin

Since no eukaryotic single-lobed transferrin is known [33], it is reasonable to think that the emergence and persistence of a bilobal structure offer substantial advantages to the organism that uses transferrin for iron transport. However, the nature of the advantages has not been confirmed. One hypothesis is that the bilobal protein resists loss through glomerular filtration in the kidney [34]. However, this hypothesis has been questioned since the bilobal structure may have evolved before the filtration kidney appeared [35]. It has been reported that the C-lobe of full-length transferrin binds iron with four times higher affinity than the isolated C-lobe at pH 7.4 [36]. This affinity becomes 25 times at pH 6.7. Iron release from the N-terminal lobe occurs in the pH range from 6 to 4 compared with 4 to 2.5 for native lactoferrin. These results also support the idea that the more facile iron release from the half-molecule (N-terminal lobe only) compared to the full-length protein is due to the absence of stabilizing interactions between N-terminal and C-terminal halves [10]. It appears that the efficiency of iron release from the C-lobe of native transferrin is impaired by stabilizing interactions of the lobes with each other that retard the release of iron. However, the binding of the transferrin receptor will overcome this problem. Thus, the bilobal structure is favored during evolution so that iron will be released from transferrin when it is complexed with the transferrin receptor.

5. Kinetics of iron release from bilobal transferrin

Although the members of the transferrin family have essentially the same fold due to the high degree of sequence identity, individual transferrin differs in their iron-binding property [37, 38]. Mechanism of iron release from each lobe differs mainly due to the differences in second shell residues [23]. *In vitro* studies with purified transferrin have shown that the iron-loaded protein releases iron as the pH is lowered [39]. Iron-loaded human serum transferrin releases iron over a pH range of 6.5 to 4, whereas the iron release from lactoferrin occurs in the pH range of 4 to 2.5 [40]. Hen serum transferrin releases the first iron in the pH range of 6.5 to 5.2 [41]. Under similar conditions, human serum transferrin 6.0 to 5.5. The loss of the remaining iron from hen serum transferrin C-lobe occurs over a pH range of 5.2 to 4. Studies performed in the absence of transferrin receptor indicate that 96% of the time, iron is released from N-lobe followed by slow release from C-lobe [42]. Also, there is cooperativity among the two lobes in the absence of transferrin receptors [43]. Thus, the iron release from the N-lobe is sensitive to the C-lobe. On the other hand, iron is released from the C-lobe 65% of the time in the presence of the transferrin receptor.

5.1 Iron release from C-lobe

Iron release from the C-lobe of transferrin is very slow and unaffected by N-lobe [44]. C-lobe has a triad of Lys534-Arg632-Asp634 that controls the iron release in the absence of the transferrin receptor [45, 46]. Lys534 and Arg632 in the C-lobe may share a H-bond that is stabilized by Asp634. Thus, the protonation of Asp634 will trigger the iron release. However, in the structure of pig transferrin, the Lys and Arg are too far away (~4.1 Å apart) to share a H-bond [47]. However, mutation of Lys/Arg to Ala severely retards iron release from C-lobe [48]. Iron release from the C-lobe in the presence of transferrin receptor proceeds via a different mechanism and is 7-10 fold faster than that in the absence of transferrin receptor. Recent studies have shown that the iron release from the C-lobe is dictated by His349 [49]. Based on the cryo-EM, it was suggested that a pair of hydrophobic residues (Trp641 and Phe760) interact with His349 and stimulates iron release by stabilizing the apo-transferrin/transferrin receptor complex [50]. The role of His349 in the iron release has been demonstrated by mutating His349 to Ala. In this H349A mutant, the iron release from C-love is reduced by 12 fold.

5.2 Iron release from N-lobe

In the absence of transferrin receptors, the iron release from N-lobe is controlled by the protonation of a pair of Lys. These two Lys residues are 3 Å apart and share a H-bond [51]. When the pH is reduced, protonation of one of the Lys residues causes the positively charged Lys residues to repeal each other (moving at least 9 Å apart) [52]. This repulsion triggers a cleft opening as well as the release of iron [53]. Mutations of any of these to two Lys to either Glu or Ala drastically slowed the rate of iron release [54]. The release of iron from the N-lobe is further facilitated by the binding of anions to Arg143 [55].

6. Stabilization of apo-transferrin/transferrin receptor complex

The return of apo-transferrin to the cell surface is a distinctive feature of the endocytic cycle. As revealed by the apo-transferrin structure, the N-lobe is stabilized by a salt bridge between Asp240 and Arg678 [56]. Additionally, the PRKP loop (residues 142-145) is connected to the bridge by a disulfide bond (between Cys137 and Cys331). In the apo-transferrin structure, the movement of the PRKP loop and the disulfide bond brings the bridge closer to the protease-like domain of the transferrin receptor to possibly further stabilize the apo-conformation in a pH-dependent manner.

7. Biological function of transferrin

Transferrin delivers the iron to various cells after binding to the transferrin receptor on the cell surface. The transferrin-transferrin receptor complex is then transported into the cell by receptor-mediated endocytosis. The iron is released from transferrin at low pH (e.g., endosomal pH). The transferrin-transferrin receptor complex will then be transported back to the cell surface, ready for another round of Fe uptake and release. This process can turn over roughly a million atoms of iron per cell per minute in active reticulocytes [57]. It is well known that Fe(III) salts are highly susceptible to hydrolysis at neutral pH to produce insoluble ferric hydroxide. Thus, the concentration of free Fe(III) in physiological fluids will be very low (~10^{-18} M). However, the daily turnover of hemoglobin iron is ~30 mg (~10^{-4} M). Thus, there is a need for a high-affinity iron-binding protein, like transferrin. By binding iron upon its release into the bloodstream, serum transferrin prevents the hydrolysis and precipitation of iron [58]. Thus, it increases the solubility of iron in the blood to the micromolar level and consequently increases its bioavailability [59]. Transferrin is usually ~30% saturated with iron with ~27% diferric transferrin, 23% monoferric transferrin (N-lobe), 11% monoferric transferrin (C-lobe), and 40% apo-transferrin [60, 61]. During increasing iron overload, the empty iron binding sites in transferrin are occupied, and thus, iron toxicity is not overserved until transferrin has been saturated with iron. Serum transferrin also inhibits the reduction of Fe(III) to Fe(II), which may lead to iron toxicity via the formation of reactive oxygen species. By having a very high affinity for Fe(III), transferrin can prevent the uptake of Fe(III) by pathogenic microorganisms. The most important role of transferrin is in the transport of iron among the site of absorption (intestinal mucosal cells), utilization (immature erythroid cells), storage (liver), and hemoglobin degradation. Thus, transferrin plays a vital and central role in iron metabolism (**Figure 9**). Although transferrin has a high molecular weight and binds only two iron ions, it is relatively efficient since it is used in many cycles of iron transport. Transferrin is recycled more than 10 times a day to supply the 20-30 mg irons needed for over 2 million erythrocytes produced every second by the bone marrow. Although iron bound to transferrin is <0.1% (4 mg) of total body iron, it constitutes the most critical iron pool with the highest turnover rate (25 mg per day) [62, 63]. It has a relatively longer half-life of 8-10 days *in vivo*. It has been suggested that plasma aluminum, when it binds to transferrin, may lead to anemia since aluminum will enter the iron distribution pathway [64].

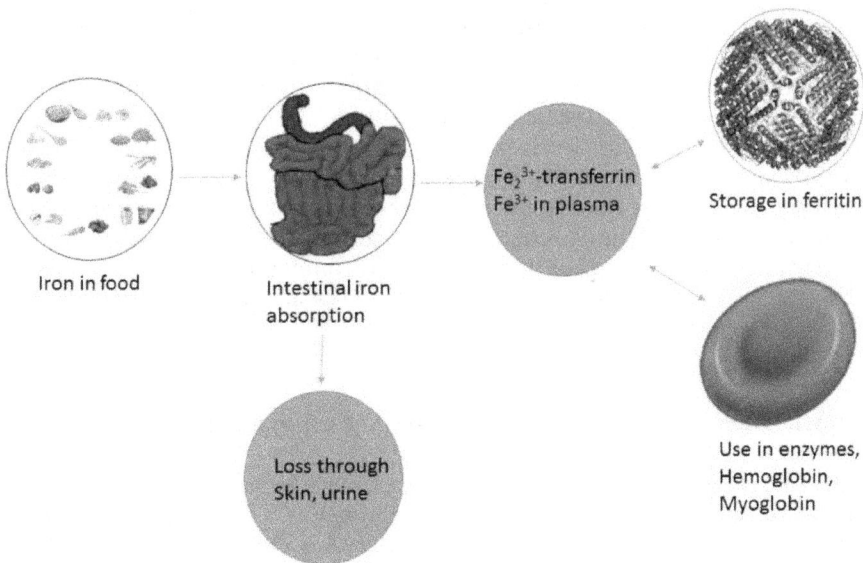

Iron in food

Intestinal iron
absorption

Fe$_2^{3+}$-transferrin
Fe^{3+} in plasma

Storage in ferritin

Loss through
Skin, urine

Use in enzymes,
Hemoglobin,
Myoglobin

Figure 9.
A simple diagram of the iron homeostasis in human.

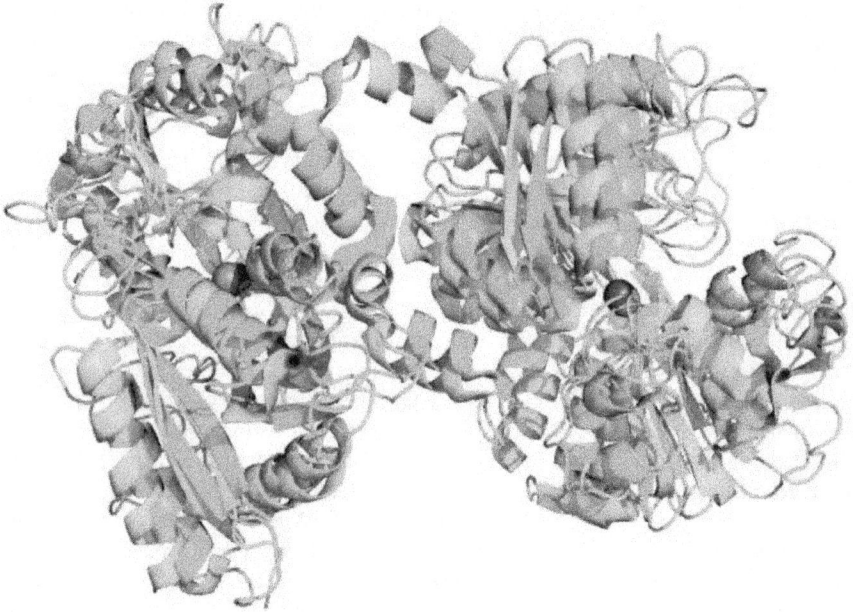

Figure 10.
Superimposition of structure of transferrin (green) and lactoferrin (cyan). Iron bound to transferrin and lactoferrin is shown as brown spheres.

In contrast, lactoferrin possesses various biological properties, including antioxidants, antimicrobial and anti-inflammatory activities [65]. Lactoferrin's affinity for iron is very high (double that of transferrin). This high iron affinity partly determines its function. Superimposition of the structure of transferrin and lactoferrin indicates that the lactoferrin has a relatively closed conformation compared to transferrin (**Figure 10**). This closed conformation may explain the higher iron affinity of lactoferrin since once iron is sequestered, it cannot escape. The high affinity of lactoferrin for iron will also enable it to deprive microorganisms of essential metals for growth. However, iron is a crucial nutrient for pathogenic microorganisms which require iron to survive and replicate. Thus, lactoferrin is considered to form part of the immune system since it deprives the pathogenic microorganisms of iron and combats the infection they cause [66]. While the apo-lactoferrin inhibits the growth of a large number of pathogenic bacteria, holo-lactoferrin shows significantly lower inhibition towards these pathogens.

8. Iron sequestration from transferrin by *N. Meningitidis*

N. Meningitidis is a pathogenic bacterium responsible for bacterial meningitis. It acquires iron from transferrin during infection through a transferrin receptor system composed of two proteins TbpA and TbpB [67]. TbpA is a 100 kDa TonB-dependent outer membrane protein required for iron uptake [68]. It can serve as a channel for iron transport across the outer membrane. TbpB is a bilobal protein (60-80 kDa) required for colonization in the host [69]. This protein extends from the outer membrane into the host milieu to interact with transferrin to initiate iron acquisition [70]. Together, these proteins sequester iron from transferrin. TbpA binds both holo- and apo-forms of transferrin [71]. However, TbpB binds only the holo-transferrin.

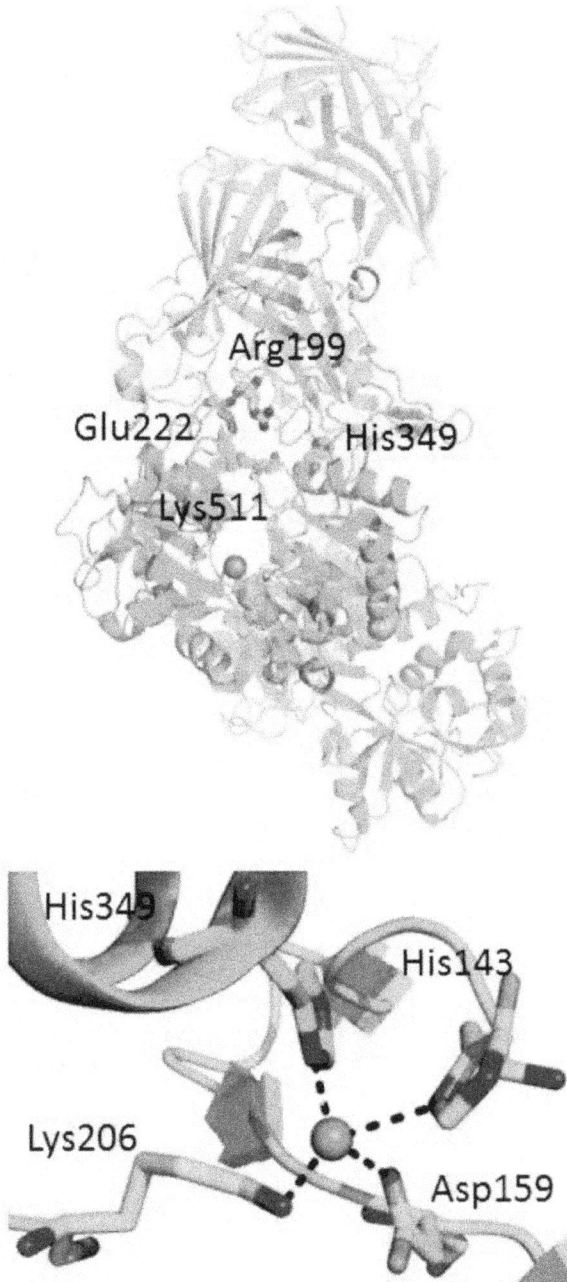

Figure 11.
Structure of transferrin (green) and transferrin binding protein TbpB (cyan) complex (top: The brown sphere is transferrin-bound iron). Interaction of His349 in transferrin with various residues in TbpB via water (sphere, bottom).

The crystal structure of TbpB with and without human transferrin has been reported [72]. Both TbpA and TbpB bind to transferrin C-lobe [73]. Within this structure, several amino acids residues of transferrin (His349, Lys511) and TbpB (Arg199, Glu222) are buried in the binding interface (**Figure 11**, top). His349 residue of transferrin interacts with TbpB through a tetrahedrally coordinated water involved in H-bonding with His143, Asp159, and Lys206 from TbpB (**Figure 11**, bottom). Protonation of His349 will cause an electrostatic repulsion with Lys511 leading to a conformational change which results in an open conformation of transferrin. Thus, His349 in human transferrin can act as a pH-inducible switch for iron release in the presence of the transferrin receptor [19, 49]. Structure-based pK_a prediction suggests that TbpB binding to human transferrin leads to a reduction of the estimated pK_a of human transferrin His349 from 6.2 (unbound form) to 1.9 (bound to TbpB) [74]. Thus, the binding of TbpB to transferrin may prevent His349 from becoming protonated and stabilize the holo C-lobe of transferrin. Therefore, TbpB does not initiate the opening of the human transferrin holo C-lobe that results in the iron release. Thus, the first step of iron acquisition by TbpB will involve binding to iron-loaded transferrin inside the host and maintaining the iron-loaded form until its delivery to TbpA in the second step.

Although the human Transferrin receptor interacts with transferrin via both N and C lobes, TbpB interacts with only the C-lobe of transferrin. Also, the human transferrin receptor binds to both apo- and holo-forms of transferrin, while TbpB binds specifically to holo-transferrin [75]. This is because the holo-form of C-lobe with a closed conformation will allow more effective docking of C-lobe of transferrin onto TbpB. However, the apo form of transferrin with an open structure will drastically reduce the binding interface between TbpB and transferrin. Unlike human transferrin receptor, TbpB interacts with a loop (residues 496-515) of human transferrin. Variation of this loop is observed among mammalian transferrins. This variation in the TbpB recognition site on transferrin seems to act as the barrier for cross-species specificity between TbpB and transferrin. Finally, bacterial transferrin binding protein (TbpB) competes with the human transferrin receptor for transferrin/iron. TbpB binding site on human transferrin partially overlaps with the transferrin receptor binding site. This overlapping binding site of the human transferrin receptor and pathogenic transferrin binding protein (e.g., TbpB) allows the pathogens to circumvent the mutation of transferrin.

9. Conclusion

This chapter highlights the role of transferrin in iron metabolism, including the iron-binding by transferrin, transferrin receptor-mediated endocytosis of transferrin/transferrin-receptor complex, and consequent iron release, etc. Some of the released iron will be stored in ferritin, while the other part will be used by various proteins and enzymes constituting a pathway of iron regulation. This iron-binding and regulation by transferrin are critical considering the toxicity of iron. How the differences (amino acids, structure, iron-binding affinity, etc.) between serum transferrin and lactoferrin dictates their biological functions have been highlighted. Finally, the competition between the human transferrin receptor and bacterial transferrin binding protein (e.g., TbpB) for getting iron from transferrin has also been discussed. Understanding the interaction of transferrin with other proteins (e.g., transferrin binding protein from various pathogenic bacteria) may lead to drug development. Overall, elucidation of the role of transferrin in iron metabolism will help understanding iron-related diseases and improve treatment.

Author details

Nitai Charan Giri
Central Institute of Petrochemical Engineering and Technology, Raipur, India

*Address all correspondence to: nitaigiri@gmail.com

IntechOpen

References

[1] Sarkar J, Potdar AA, Saidel GM. Whole-body iron transport and metabolism: Mechanistic, multi-scale model to improve treatment of anemia in chronic kidney disease. PLoS Comput Biol. 2018;14(4):e1006060. DOI: 10.1371/journal.pcbi.1006060.

[2] Yiannikourides A, Latunde-Dada GO. A Short Review of Iron Metabolism and Pathophysiology of Iron Disorders. Medicines (Basel). 2019;6(3):85. DOI: 10.3390/medicines6030085.

[3] Harris DC. & Aisen, P. In Iron Carriers and Iron Proteins (Loehr, T. M., ed.), VCH Publishers, Inc., New York; 1989. p. 239-352.

[4] Bertini I, Gray HB, Stiefel EI, Valentine JS. Biological Inorganic Chemistry: Structure and Reactivity; University Science Books: Mill Valley, CA, USA; 2007. 628 p.

[5] Baker EN, Lindley PF. New perspectives on the structure and function of transferrins. J. Inorg. Biochem. 1992;47:147-160. DOI: 10.1016/0162-0134(92)84061-q

[6] Aisen P. in: A. Sigel, H. Sigel (Eds.), Metals in Biological Systems, Marcel Dekker, New York, 1998, p. 585-665.

[7] . Mecklenburg SL, Donohoe RJ, Olah GA. J. Mol. Biol. 1977;270:739-750. DOI: 10.1006/jmbi.1997.1126

[8] Li H, Sadler PJ, Sun H. Rationalization of the strength of metal binding to human serum transferrin. Eur. J. Biochem. 1996;242:387-393. DOI: 10.1111/j.1432-1033.1996.0387r.x

[9] Wally J, Halbrooks PJ, Vonrhein C, Rould MA, Everse SJ, Mason AB, Buchanan SK. The crystal structure of iron-free human serum transferrin provides insight into inter-lobe communication and receptor binding. J. Biol. Chem. 2006;281:24934-24944. DOI: 10.1074/jbc.M604592200

[10] Lee D, and J Goodfellow. The pH-induced release of iron fromtransferrin investigated with a continuum electrostatic model. Biophys. J. 1998;74:2747-2759. DOI: 10.1016/S0006-3495(98)77983-4

[11] Lambert LA, Perri H, Halbrooks PJ, Mason AB. Evolution of the transferrin family: Conservation of residues associated with iron and anion binding. Comp. Biochem. Physiol. B Biochem. Mol. Biol. 2005;142:129-141. DOI: 10.1016/j.cbpb.2005.07.007

[12] Lambert LA, Perri H, Meehan TJ. Evolution of duplications in the transferrin family of proteins. Comp. Biochem. Physiol. B Biochem. Mol. Biol. 2005;140:11-25. DOI: 10.1016/j.cbpc.2004.09.012

[13] Tinoco AD, Peterson CW, Lucchese B, Doyle RP, Valentine AM. On the evolutionary significance and metal-binding characteristics of a monolobal transferrin from Ciona intestinalis. Proc. Natl. Acad. Sci. USA 2008, 105, 3268-3273. DOI: 10.1073/pnas.0705037105

[14] Bates GW, Workman EF Jr, Schlabach MR. Does transferrin exhibit ferroxidase activity? Biochem. Biophys. Res. Commun. 1973;50:84-90. DOI: 10.1016/0006-291X(73)91067-X

[15] Kojima N, Bates GW. The formation of Fe3+-transferrin-CO3 2- via the binding and oxidation of Fe2+. J. Biol.

Chem. 1981;256:12034-12039.
DOI: 10.1016/S0021-9258(18)43229-2

[16] Grossmann JG, Crawley JB, Strange RW, Patel KJ, Murphy LM, Neu M, Evans RW, Hasnain SS. The nature of ligand-induced conformational change in transferrin in solution. An investigation using X-ray scattering, XAFS and site-directed mutants. J. Mol. Biol. 1998;279(2):461-72. DOI: 10.1006/jmbi.1998.1787.

[17] Daniels TR, Delgado T, Rodriguez JA, Helguera G, Penichet ML. The transferrin receptor part I: Biology and targeting with cytotoxic antibodies for the treatment of cancer. Clin. Immunol. 2006;121:144-158. DOI: 10.1016/j.clim.2006.06.010

[18] Cheng Y, Zak O, Aisen P, Harrison SC, Walz T. Structure of the human transferrin receptor-transferrin complex. Cell 2004;116: 565-576. DOI: 10.1016/s0092-8674(04)00130-8

[19] Eckenroth BE, Steere AN, Chasteen ND, Everse SJ, Mason AB. How the binding of human transferrin primes the transferrin receptor potentiating iron release at endosomal pH. Proc. Natl. Acad. Sci. USA 2011;108:13089-13094. DOI: 10.1073/pnas.1105786108

[20] Leverence R, Mason AB, Kaltashov IA. Noncanonical interactions between serum transferrin and transferrin receptor evaluated with electrospray ionization mass spectrometry. Proc. Natl. Acad. Sci. USA 2010;107:8123-8128. DOI: 10.1073/pnas.0914898107

[21] Conner SD, Schmid SL. Differential requirements for AP-2 in clathrin-mediated endocytosis. J. Cell Biol. 2003;162:773-779. DOI: 10.1083/jcb.200304069

[22] Ohgami RS, Campagna DR, Greer EL, Antiochos B, McDonald A, Chen J, Sharp JJ, Fujiwara Y, Barker JE, Fleming MD. Identification of a ferrireductase required for e_cient transferrin-dependent iron uptake in erythroid cells. Nat. Genet. 2005;37:1264-1269. DOI: 10.1038/ng1658

[23] Steere AN, Byrne SL, Chasteen ND, Mason AB. Kinetics of iron release from transferrin bound to the transferrin receptor at endosomal pH. Biochim. Biophys. Acta 2012;1820:326-333. DOI: 10.1016/j.bbagen.2011.06.003

[24] Kakhlon O, Cabantchik ZI. The labile iron pool: Characterization, measurement, and participation in cellular processes. Free Radic. Biol. Med. 2002;33:1037-1046. DOI: 10.1016/s0891-5849(02)01006-7

[25] Kruszewski M. Labile iron pool: The main determinant of cellular response to oxidative stress. Mutat. Res. Fundam. Mol. Mech. Mutag. 2003;531:81-92. DOI: 10.1016/j.mrfmmm.2003.08.004

[26] Breuer W, Shvartsman M, Cabantchik ZI. Intracellular labile iron. Int. J. Biochem. Cell Biol. 2008;40:350-354. DOI: 10.1016/j.biocel.2007.03.010

[27] Cabantchik ZI. Labile iron in cells and body fluids: Physiology, pathology, and pharmacology. Front. Pharmacol. 2014;5:45. DOI: 10.3389/fphar.2014.00045

[28] Kraiter DC, Zak O, Aisen P, Crumbliss AL. A determination of the reduction potentials for diferric and C- and N-lobe monoferric transferrins at endosomal pH (5.8). Inorg. Chem. 1998;37:964-968. DOI: 10.1021/ic970644g

[29] Dhungana S, Taboy CH, Zak O, Larvie M, Crumbliss AL, Aisen P. Redox properties of human transferrin bound to

its receptor. Biochemistry 2004;43: 205-209. DOI: 10.1021/bi0353631

[30] Bou-Abdallah F. Does Iron Release from Transferrin Involve a Reductive Process? Bioenerg. Open Access 2012;1:1000e111.

[31] Harris WR. Thermodynamic binding constants of the zinc-human serum transferrin complex. Biochemistry 1983;22:3920-3926. DOI: 10.1021/bi00285a030

[32] Levina A, Lay PA. Transferrin cycle and clinical roles of citrate and ascorbate in improved iron metabolism. ACS Chem. Biol. 2019;14:893-900. DOI: 10.1021/acschembio.8b01100

[33] Zak O, Aisen P. Iron release from transferrin, its C-Lobe, and their complexes with transferrin receptor: Presence of N-lobe accelerates release from C-lobe at endosomal pH. Biochemistry. 2003;42:12330-12334. DOI: 10.1021/bi034991f

[34] Williams J. The evolution of transferrin. Trends Biochem. Sci. 1982;7:394-397. DOI: 10.1016/0968-0004(82)90183-9

[35] . Gasdaska J R, Law J H, Bender C J, Aisen P. Cockroach transferrin closely resembles vertebrate transferrins in its metal ion-binding properties: A spectroscopic study. J. Inorg. Biochem. 1996;64:247-258. DOI: 10.1016/S0162-0134(96)00052-9

[36] Zak O, Aisen P. Preparation and properties of a single-sited fragment from the C-terminal domain of human transferrin. Biochim. Biophys. Acta 1985;829:348-353. DOI: 10.1016/0167-4838(85)90243-2

[37] Day CL, Stowell KM, Baker EN, Tweedie JW. Studies of the N-terminal half of human lactoferrin produced from the cloned cDNA demonstrate that interlobe interactions modulate iron release, J. Biol. Chem. 1992;267:13857-13862. DOI: 10/1016/S0021-9258 (19)49647-6

[38] Abdallah FB, El Hage Chahine JM. Transferrins, the mechanism of iron release by ovotransferrin, Eur. J. Biochem. 1999;263:912-920. DOI: 10.1046/j.1432-1327.1999.00596.x

[39] Thakurta PG, Choudhury D, Dasgupta R, et al. Tertiary structural changes associated with iron binding and release in hen serum transferrin: a crystallographic and spectroscopic study. Biochem Biophys Res Commun. 2004; 316:1124-1131. DOI: 10.1016/j.bbrc.2004.02.165

[40] Mazurier J, G. Spik. Comparative study of the iron-binding properties of human transferrins. I. Complete and sequential iron saturation and desaturation of the lactotransferrin. Biochim. Biophys. Acta 1980;629: 399-408. DOI: 10.1016/0304-4165 (80)90112-9

[41] Peterson NA, Anderson BF, Jameson GB, Tweedie JW, Baker EN. Crystal structure and iron-binding properties of the R210K mutant of the N-lobe of human lactoferrin: implications for iron release from transferrins, Biochemistry. 2000;39:6625-6633. DOI: 10.1021/bi0001224

[42] Byrne SL, Chasteen ND, Steere AN, Mason AB. The unique kinetics of iron release from transferrin: The role of receptor, lobe-lobe interactions, and salt at endosomal pH. J Mol Biol 2010;396: 130-140. DOI: 10.1016/j.jmb.2009.11.023

[43] . Byrne SL, Mason AB. Human serum transferrin: A tale of two lobes. Urea gel and steady state fluorescence analysis of

recombinant transferrins as a function of pH, time, and the soluble portion of the transferrin receptor. J Biol Inorg Chem. 2009;14:771-781. DOI: 10.1007/s00775-009-0491-y

[44] . Bali PK, Aisen P. Receptor-induced switch in site-site cooperativity during iron release by transferrin. Biochemistry. 1992;31:3963-3967. DOI: 10.1021/bi00131a011

[45] . Halbrooks PJ, et al. Composition of pH sensitive triad in C-lobe of human serum transferrin. Comparison to sequences of ovotransferrin and lactoferrin provides insight into functional differences in iron release. Biochemistry. 2005;44:15451-15460. DOI: 10.1021/bi0518693

[46] James NG, et al. Inequivalent contribution of the five tryptophan residues in the C-lobe of human serum transferrin to the fluorescence increase when iron is released. Biochemistry. 2009; 48:2858-2867. DOI: 10.1021/bi8022834

[47] Hall DR, Hadden JM, Leonard GA, Bailey S, Neu M, Winn M, Lindley PF. The crystal and molecular structures of diferric porcine and rabbit serum transferrins at resolutions of 2.15 and 2.60 Å, respectively. Acta. Crystallogr. D Biol. Crystallogr. 2002; 58:70-80. DOI: 10.1107/s0907444901017309

[48] Halbrooks PJ, He QY, Briggs SK, Everse SJ, Smith VC, MacGillivray RT, Mason AB. Investigation of the mechanism of iron release from the C-lobe of human serum transferrin: mutational analysis of the role of a pH sensitive triad. Biochemistry. 2003; 42:3701-3707. DOI: 10.1021/bi027071q

[49] . Steere AN, et al. Evidence that His349 acts as a pH-inducible switch to accelerate receptor-mediated iron release

from the C-lobe of human transferrin. J Biol Inorg Chem. 2010;15:1341-1352. DOI: 10.1007/s00775-010-0694-2

[50] Giannetti AM, et al. The molecular mechanism for receptor-stimulated iron release from the plasma iron transport protein transferrin. Structure. 2005;13:1613-1623. DOI: 10.1016/j.str.2005.07.016

[51] . Dewan JC, Mikami B, Hirose M, Sacchettini JC. Structural evidence for a pH-sensitive dilysine trigger in the hen ovotransferrin N-lobe: implications for transferrin iron release. Biochemistry. 1993; 32:11963-11968. DOI: 10.1021/bi00096a004

[52] Jeffrey PD, Bewley MC, MacGillivray RT, Mason AB, Woodworth RC, Baker EN. Ligand-induced conformational change in transferrins: crystal structure of the open form of the N-terminal half-molecule of human transferrin. Biochemistry. 1998; 37:13978-13986. DOI: 10.1021/bi9812064

[53] He QY, Mason AB, Tam BM, MacGillivray RTA, Woodworth RC. Dual role of Lys206-Lys296 interaction in human transferrin N-lobe: Iron-release trigger and anion-binding site. Biochemistry. 1999;38:9704-9711. DOI: 10.1021/bi990134t

[54] He Q-Y, Mason AB. Molecular aspects of release of iron from transferrin. D.M. Templeton; New York: 2002

[55] Byrne SL, Steere AN, Chasteen ND, Mason AB. Identification of a kinetically significant anion binding (KISAB) site in the N-lobe of human serum transferrin. Biochemistry. 2010;49:4200-4207. DOI: 10.1021/bi1003519

[56] Wally J, et al. The crystal structure of iron-free human serum transferrin provides insight into inter-lobe

communication and receptor binding. J Biol Chem. 2006;281:24934-24944. DOI: 10.1074/jbc.M604592200

[57] Aisen P, Listowsky I. Iron transport and storage proteins. Annu. Rev. Biochem. 1980;49:357. DOI: 10.1146/annurev.bi.49.070180.002041

[58] Stefánsson A. Iron(III) Hydrolysis and Solubility at 25 °C. Environ. Sci. Technol. 2007; 41:6117-6123. DOI: 10.1021/es070174h

[59] Aisen P, Leibman A, Zweier J. Stoichiometric and site characteristics of the binding of iron to human transferrin. J. Biol. Chem. 1978;253:1930-1937.

[60] G. C. Ford et al. Ferritin: design and formation of an iron-storage molecule. Philos. Trans. Roy. Soc. Land. B. 1984; 304:551-565. DOI: 10.1098/rstb.1984.0046

[61] Williams J, Moreton K. The distribution of iron between the metal-binding sites of transferrin in human serum. Biochem J. 1980;185:483-488. DOI: 10.1042/bj1850483

[62] Gkouvatsos K, Papanikolaou G, Pantopoulos K. Regulation of iron transport and the role of transferrin. Biochim. Biophys. Acta. 2012;1820: 188-202. DOI: 10.1016/j.bbagen.2011.10.013

[63] Lambert LA. Molecular evolution of the transferrin family and associated receptors. Biochim. Biophys. Acta. 2012;1820:242-253. DOI: 10.1016/j.bbagen.2011.06.002

[64] El Hage Chahine JM, Hemadi M, Ha-Duong NT. Uptake and release of metal ions by transferrin and interaction with receptor 1. Biochim. Biophys. Acta 2012, 1820, 334-347. DOI: 10.1016/j.bbagen.2011.07.008

[65] Baker EN, Baker HM. A structural framework for understanding the multifunctional character of lactoferrin. Biochimie. 2009;91:3-10. DOI: 10.1016/j.biochi.2008.05.006

[66] Chung MCM. Structure and function of transferrin. Biochem Edu. 1984;12:146-54. DOI: 10.1016/0307-4412%2884%2990118-3

[67] Gray-Owen SD, Schryvers AB. Bacterial transferrin and lactoferrin receptors. Trends Microbiol. 1996;4:185-191. DOI: 10.1016/0966-842x(96)10025-1

[68] Cornelissen CN, et al. The transferrin receptor expressed by gonococcal strain FA1090 is required for the experimental infection of human male volunteers. Mol. Microbiol. 1998;27:611-616. DOI: 10.1046/j.1365-2958.1998.00710.x

[69] Baltes N, Hennig-Pauka I, Gerlach GF. Both transferrin binding proteins are irulence factors in *Actinobacillus pleuropneumoniae* serotype 7 infection FEMS Microbiol. Lett. 2002;209:283-287. DOI: 10.1111/j.1574-6968.2002.tb11145.x

[70] Moraes TF, Yu RH, Strynadka NC, Schryvers AB. Insights into the Bacterial Transferrin Receptor: The Structure of Transferrin-Binding Protein B from Actinobacillus pleuropneumoniae Mol. Cell 2009;35:523-533. DOI: 10.1016/j.molcel.2009.06.029

[71] Retzer MD, Yu R, Zhang Y, Gonzalez GC, Schryvers AB. Discrimination between apo and iron-loaded forms of transferrin by transferrin binding protein B and its N-terminal subfragment Microb. Pathog. 1998;25:175-180. DOI: 10.1006/mpat.1998.0226

[72] Calmettes C, Alcantara J, Yu RH, et al. The structural basis of transferrin

sequestration by transferrin-binding protein B. Nat Struct Mol Biol. 2012;19:358-360. DOI:10.1038/nsmb.2251

[73] Alcantara, J., Yu, R.H. & Schryvers, A.B. The region of human transferrin involved in binding to the bacterial transferrin receptors is localised in the C-lobe. Mol. Microbiol. 1993;8:1135-1143. DOI: 10.1111/j.1365-2958.1993.tb01658.x

[74] Li H, Robertson AD, Jensen JH. Very fast empirical prediction and rationalization of protein pKa values. Proteins. 2005;61:704-721. DOI: 10.1002/prot.20660

[75] Schryvers AB, Gonzalez GC. Receptors for transferrin in pathogenic bacteria are specific for the host's protein. Can. J. Microbiol. 1990;36:145-147 (1990). DOI: 10.1139/m90-026

Chapter 4

Hepcidin

Safa A. Faraj and Naeem M. Al-Abedy

Abstract

The hepcidin is antimicrobial peptide has antimicrobial effects discover before more than a thousand years; it has a great role in iron metabolism and innate immunity. Hepcidin is a regulator of iron homeostasis. Its production is increased by iron excess and inflammation and decreased by hypoxia and anemia. Iron-loading anemias are diseases in which hepcidin is controlled by ineffective erythropoiesis and concurrent iron overload impacts. Hepcidin reacts with ferroportin. The ferroportin is found in spleen, duodenum, placenta, if the ferroportin decrease, it results in the reduced iron intake and macrophage release of iron, and using the iron which stores in the liver. Gene of human hepcidin is carried out by chromosome 19q13.1. It consists of (2637) nucleated base. HAMP gene was founded in the liver cells, in brain, trachea, heart, tonsils, and lung. Changing in the HAMP gene will produce a change in hepcidin function. The hepcidin is made many stimulators are included opposing effects exerted by pathological and physiological conditions. Hepcidin is essential for iron metabolism, understanding stricter and genetic base of hepcidin is crucial step to know iron behavior and reactions to many health statuses.

Keywords: hepcidin, iron, HAMP gene

1. Introduction

The hepcidin is antimicrobial peptide has antimicrobial effects discover before more than a thousand years; it has a great role in iron metabolism and innate immunity [1]. It's a peptide hormone produced by the liver that acts as an iron regulator. Hepcidin is an iron homeostasis regulator. Iron deficiency and inflammation boost its production, while hypoxia and anemia diminish it. Hepcidin prevents iron from duodenal enterocytes absorbing dietary iron, macrophages recycling iron from senescent erythrocytes, and iron-storing hepatocytes from entering the bloodstream. Hepcidin is controlled by inefficient erythropoiesis and concurrent iron overload consequences in iron-loading anemias [2].

Human urine and blood, particularly plasma after filtration, were used to isolate hepcidin. Macrophages, adipocytes, neutrophils, lymphocytes, kidney cells, and -cells all make hepcidin. The studies experiment on mice used for determination hepcidin regulation, expression, function, and structure. Severe iron overload is occurring due to the gene responsible with hepcidin production; the gen has the role of iron function. However, decreased and iron increased hepcidin expression in transgenic animals. Hepcidin serves a variety of purposes, including inflammation, hypoxia, and iron storage [3].

IntechOpen

Ferroportin reacts with hepcidin. The ferroportin is located in the spleen, duodenum, and placenta; a decrease in ferroportin results in reduced iron intake and iron release by macrophages, as well as the use of iron stored in the liver [4].

2. The *HAMP* gene and structure of hepcidin

The human hepcidin gene is located on chromosome 19q13.1. It is made up of 263 nucleated bases. The HAMP gene was discovered in liver cells, brain cells, trachea, heart, tonsils, and lung cells [5].

Hepcidin comes in three forms: 25 aa, 22 aa, and 20 aa peptide. The HAMP gene encodes preprohepcidin, which has 84 amino acids. The structure of hepcidin25 (**Figure 1**), which consists of (8) cysteine linked by a disulfide bond, is detected in urine, while 25 and 20 are found in human serum. The structure of hepcidin is studied using NMR spectroscopy; it has four disulfide links [7].

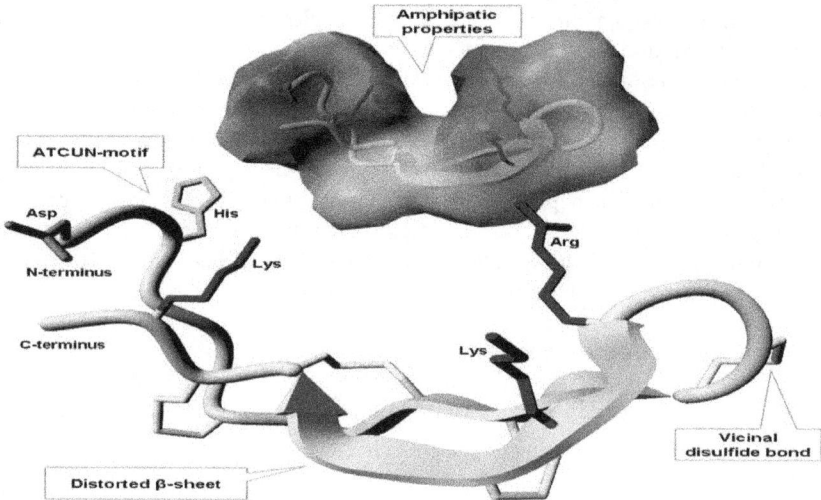

Figure 1.
Molecule structure of human synthetic hepcidin-25. Background: Hepcidin-25. Front: Showing the general structure of hepcidin-25. Gray arrows are distorted β-sheets, and colored gray is peptide backbone. Colored yellow is a disulfide bond, blue is indicate to positive residues of lysine and arginine, red indicates to the negative residue of aspartic acid, and colored green indicates to histidine which containing amino-terminal [6].

3. Hepcidin gene regulation

Location of the HAMP gene is at 19q13 chromosome mRNA. Several genetic factors affect on iron concentration, hypoxia, inflammation, erythropoiesis, and anemia. All these factors have two pathways on the gene. The first signaling is by bone proteins and the second Janus kinase/signal related to inflammation [8].

The protein is regulated of hepcidin level depend on transferrin and interaction receptor. HFE is chanced from TfR1 [Tf-Fe3+] to promote its interaction with (TfR2).

TfR2 and HFE link with the receptor of hemojuvelin by the BMP/Son for activating HAMP gene. This reaction stimuli phosphorylation of BMP receptor, and stimulating signals into the cell. The receptor of type II activates receptor of type I, then the signal transmits to the SMAD regulatory receiver, phosphorylating (SMAD-8, SMAD-5, and SMAD-1). The activated complex transfer to the nucleus for regulating gene transcription [9]. Matriptase-2 protein and SMAD-4 is a suppressor of BMP/SMAD. HJV is reacting with Matriptase-2 and causes fragmentation. Growth hormone and erythropoietin associate with the receptor, wherever, interferon and cytokines. The hepcidin wad produced in the liver, it increases if iron gets in liver cells. Hepcidin creates and released into the blood vessels and spread all the body. It interacts with other proteins in the liver, intestines, and WBC for iron storage when the hepcidin was produced at large amounts, increases the occurrence of liver tumor or chronic or acute hypoferremia. If the hepcidin is decreased in the production, results in mutations in the hemojuvelin gene, hepcidin gene, or *transferrin receptor 2* [10].

4. *HAMP* gene mutation

Hepcidin function will be altered if the HAMP gene is altered. Exon 3 of the HAMP gene encodes proteins, and it is regarded the most critical and biggest section of the gene, containing several polymorphisms [10]. The HFE gene has more polymorphisms than the HAMP gene. There are approximately 16 forms of single nucleotide polymorphism that have been discovered in various investigations [8]. Mutations in the gene have been reported in a number of reports. People who have mutations in the HAMP gene develop juvenile hemochromatosis between the ages of 10 and 30. As the initial genetic change in the HAMP gene, microsatellite marker probes are utilized. After exchanging several amino acids in the active peptide, or replacing C78 with a tyrosine, C78T, the mutation occurs in c.233G > A at some point [11].

The mutation allows ferroportin to form bisulfite connections with hepcidin, resulting in an increase in iron absorption. C70R mutations result in cysteine bisulfite bond distortion. The arginine replaces the cysteine, which does not allow the creation of the bisulfite bridge between 3 and 6 in the hepcidin peptide. C to T substitutions were found at position (166) of the HAMP (166C-T), as well as arginine substitutions at position (56) for a halting codon (R56X), 193A to R56X. (T). Furthermore, the ferroprotein does not bind to hepcidin, resulting in the production of additional iron. In contrast, deleting guanine from exon two at location 93 causes an RNA mutation. The deletion of Met50del IVS21 from exon two causes a disruption in the active peptide's expression as well as variations in reading frames. Met50 and (IVS + 1 (G)) are suppressed by the mutation. The reading frame is lengthened as a result of this mutation. Another mutation, G71D, causes a change in amino acid 71, which lies between 3 and 4 cysteine and precludes ferroprotein binding. In sickle cell disease patients, the HFE-H63D mutation is linked to the HAMP-G71D variant, which increases iron overload [12].

The polymorphism (G to A) occurs at the +14 position of the 5'-UTR region, resulting in a new initiation codon, a new aberrant protein, and a shift in the reading frame. After the mRNA is translated, an unstable protein will be produced, which will be analyzed. The related polymorphisms NC-582A > G and NC-1010C > T in the HAMP gene create a haplotype with ferritin concentrations greater than 300 g/L [13].

HFE gene polymorphisms are frequently linked to HAMP. With iron overload, there are various mixed clinical symptoms in some clinical instances. The variations C, 582A > G and C-153C > T reduce hepcidin expression, but the peptide's mode of action remains same without transferrin saturation and increased ferritin levels. The patient who has HAMP gene mutations were cannot make hepcidin and unable to decrease iron absorption. The body organs become contain iron at large amounts such as heart and liver, and it will affect with damage. Any change in the HAMP gene could result in a faulty hepcidin protein, and it would have no effect. The accumulation of iron and ferritin in the organs contributes to the development of diseases in several organs, such as coronary artery disease, diabetes mellitus, HIV, HBV, and HCV, where reactive oxygen generates oxidative material that damages tissues. And some neurological illnesses, such as Alzheimer's, Parkinson's, and sclerosis, are linked to high levels of hepcidin in the blood [14].

5. The hepcidin clinical applications

The hepcidin is made many stimulators are included opposing effects exerted by pathological and physiological conditions. The response is usually rapid. The hepcidin production increases during few hours after inflammatory stimulation and iron administration. Several stimuli could associate with hepcidin. Such as in hepcidin production and severe ID with the inflammation [15, 16].

Several ineffective conditions, such as signals released by bone marrow and non-transfusion-dependent thalassemia. The results showed hepcidin suppression non-transfusion-dependent thalassemias other iron-loading anemias, and even in-thalassemia trait. Serum hepcidin in transfusion-dependent b-thalassemia showed increasing in blood transfusions and decreasing through inter-transfusion periods [16].

Clinically relevant conditions include CKD, RBC transfusions, iron administration, replete iron stores, TMPRSS6 variants, infections/inflammatory disorders, ineffective erythropoiesis, hypoxia, erythropoietic stimulating agent administration, chronic liver diseases, alcohol abuse, HCV, hemochromatosis-related mutations, and testosterone estrogen administration. HCV, hereditary hemochromatosis; HH, iron deficiency; IDA, RBC, transmembrane protease serine 6, matriptase-2 encoding gene; CKD, glomerular filtration rate; GFR, hepatitis C virus; HCV, hereditary hemochromatosis; IDA, RBC, transmembrane protease serine 6, and matriptase-2 encoding [17].

6. Structure and location of *HFE* gene human

HFE protein was encoded by the HFE gene in humans. The gene lies at chromosome six—6p21.3. The protein is included membrane protein such as MHC class I-type and link with beta-2 microglobulin. HFE protein regulates iron uptake by transferrin HFE protein and the transferrin receptor which composed from (343) amino acid (**Figure 2**). Many other types of proteins such as a signal peptide, transferrin receptor-binding region, and immunoglobulin molecules. HFE is prominent in small intestinal absorptive cells, epithelial cells in stomach, macrophages, and granulocytes and monocytes [19].

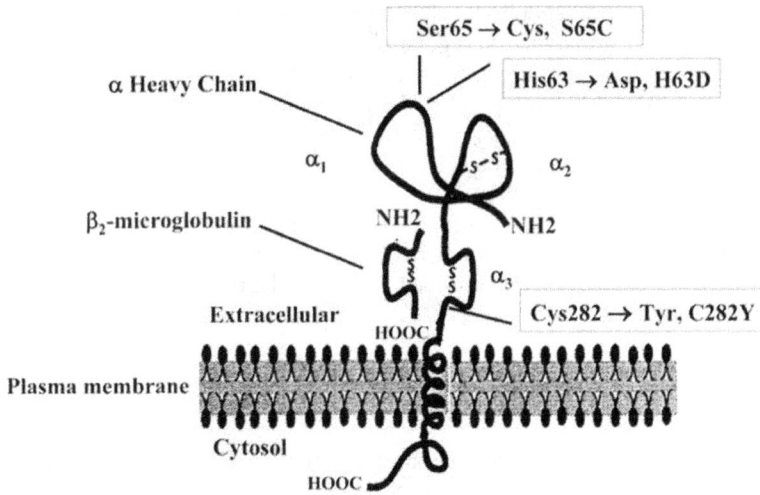

Figure 2.
The HFE gene diagram. The image was changed after getting permission from the author. Cys282 -> Tyr282 exchanging mutation of C282Y and His63 -> Asp63 exchanging mutation of H63D [18].

7. Maintaining iron homeostasis by hepcidin

Hepcidin is regulated iron absorption. Pathway of iron is showed in (**Figure 3**), FPN1link with hepcidin is results in iron retention into the cell and do not allow of iron from getting in the plasma. Hepcidin is made and store in the liver cell [20].

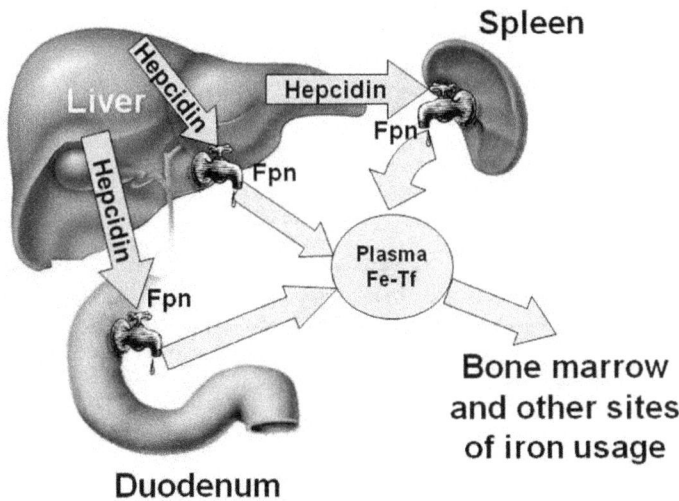

Figure 3.
Hepcidin internalization and degradation.

Ferroportin binds to hepcidin results in degradation, wherever the reaction between ferroportin and hepcidin regulates and control on iron concentration. The hepcidin regulation is a very complex mechanism and depends on many transmembrane proteins. JAK-STAT activated HAMP expression in interleukin-6 (IL6) status and inflammation-mediated response [21].

Bone-morphogenetic protein is work as a key for the regulation of *HAMP* gene through *SMAD* signaling pathway.

Also, hemojuvelin is protein made in a lever membrane cell. If the iron becomes low, the hemojuvelin is activated wherever the *sHJV* inhibits *HAMP level* through lining with *BMP*. Regulation of iron is not understood although many proteins work to iron regulation we know it such as (*TFR2*) have great role in iron regulation [22].

8. Iron regulation by hepcidin

The function of the enterocytes has absorbed the iron, the iron store in the macrophages and hepatocytes and the process is controlled by hepcidin, the hepcidin is produced in the liver. Hepcidin consists of (84) amino acid, it is undergoing for several reactions to become (60) amino acid then transfer to (25) amino acid [23].

The hepcidin is hormone consist of four disulphide bonds and 32% beta-sheet. The function of the hepcidin is control on iron efflux by ferroportin wherever; the liver secretes iron in the plasma. After secretion, the hepcidin is bound with ferroportin; wherever, ferroportin is protein on the cell surface have transferred the iron inside the cell. If the ferroportin is reduced in expression, become the intracellular iron is less. Hepcidin is absorbed iron from the food and transfers to plasma, and the

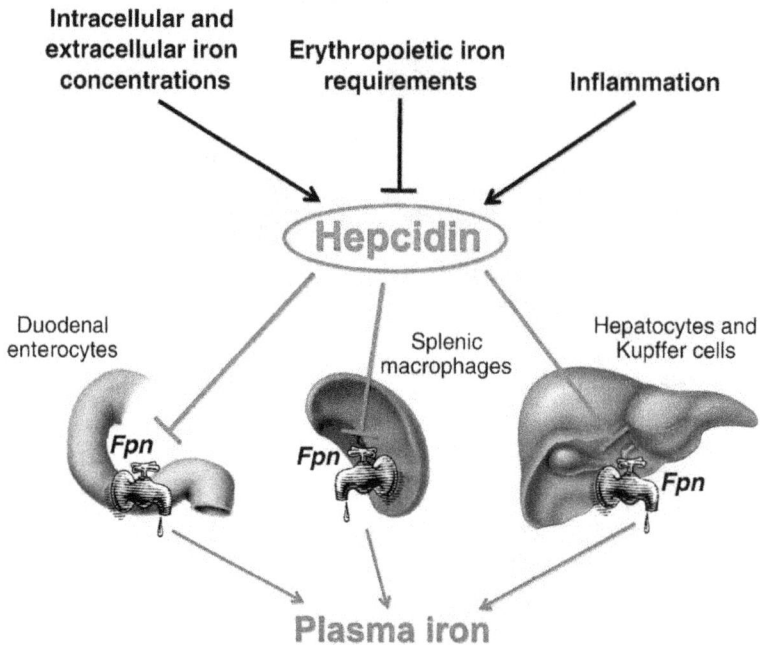

Figure 4.
Regulation of iron balance.

iron gets in the cell by binding hepcidin and ferroportin. Reduction of the ferroportin on the cell surface is mechanism unclear [9].

Figure 4 show how iron flows into plasma exclusively through the membrane channel, ferroportin. Macrophages, enterocyte's hepatocytes are the principal cell types that express ferroportin and so export iron. The duodenum, spleen, and liver, which contain these cells, are important locations for controlling iron flux (blue arrows). Hepcidin, a 25-amino-acid hepatic hormone, regulates ferroportin levels. Endocytosis and proteolysis are triggered when hepcidin binds to ferroportin, preventing iron flow (red arrows) into the plasma from ferroportin-expressing tissues. Hepcidin production rises as iron stocks rise during infection (black arrows) and falls as erythropoiesis demands more iron (red arrows) [24].

9. Conclusion

Hepcidin is essential for iron metabolism, understanding stricter and genetic base of hepcidin is crucial step to know iron behavior and reactions to many health status. This chapter highlights on hepcidin structure and genetic information as well as its relation to iron metabolism.

Author details

Safa A. Faraj[1,2]* and Naeem M. Al-Abedy[3]

1 Department of Pediatrics, College of Medicine, Wasit University, Kut, Iraq

2 Children Welfare Teaching Hospital, Baghdad Medical City, Baghdad, Iraq

3 Al-Karama Teaching Hospital, Wasit, Iraq

*Address all correspondence to: safaafaraj@uowasit.edu.iq

IntechOpen

References

[1] Mouchahoir T, Schiel JE. Development of an LC-MS/MS peptide mapping protocol for the NISTmAb. Analytical and Bioanalytical Chemistry. 2018;**410**(8):2111-2126

[2] Pigeon C, Ilyin G, Courselaud B, Leroyer P, Turlin B, Brissot P, et al. A new mouse liver-specific gene, encoding a protein homologous to human antimicrobial peptide hepcidin, is overexpressed during iron overload. Journal of Biological Chemistry. 2001;**276**(11):7811-7819

[3] Dlouhy AC, Bailey DK, Steimle BL, Parker HV, Kosman DJ. Fluorescence resonance energy transfer links membrane ferroportin, hephaestin but not ferroportin, amyloid precursor protein complex with iron efflux. Journal of Biological Chemistry. 2019;**294**(11):4202-4214

[4] Béliveau F, Tarkar A, Dion SP, Désilets A, Ghinet MG, Boudreault PL, et al. Discovery and development of TMPRSS6 inhibitors modulating hepcidin levels in human hepatocytes. Cell Chemical Biology. 2019;**26**(11):1559-1572

[5] Elnabaheen EM. Hepcidin Status among Iron Deficient Anemic Pregnant Women in Gaza strip: A Case Control Study. 2017. Available from: https://iugspace.iugaza.edu.ps/handle/20.500.12358/21572

[6] Kemna EH, Kartikasari AE, van Tits LJ, Pickkers P, Tjalsma H, Swinkels DW. Regulation of hepcidin: Insights from biochemical analyses on human serum samples. Blood Cells, Molecules, and Diseases. 2008;**40**(3):339-346

[7] Huang T, Gu W, Wang B, Zhang Y, Cui L, Yao Z, et al. Identification and expression of the hepcidin gene from brown trout (*Salmo trutta*) and functional analysis of its synthetic peptide. Fish & Shellfish Immunology. 2019;**87**:243-253

[8] Fleming RE, Sly WS. Hepcidin: A putative iron-regulatory hormone relevant to hereditary hemochromatosis and the anemia of chronic disease. Proceedings of the National Academy of Sciences. 2001;**98**(15):8160-8162

[9] Fillebeen C, Charlebois E, Wagner J, Katsarou A, Mui J, Vali H, et al. Transferrin receptor 1 controls systemic iron homeostasis by fine-tuning hepcidin expression to hepatocellular iron load. Blood. 2019;**133**(4):344-355

[10] Pandey S, Pandey SK, Shah V. Role of HAMP genetic variants on pathophysiology of iron deficiency anemia. Indian Journal of Clinical Biochemistry. 2018;**33**(4):479-482

[11] Arts HH, Eng B, Waye JS. Multiplex allele-specific PCR for simultaneous detection of H63D and C282Y HFE mutations in hereditary hemochromatosis. Journal of Applied Laboratory Medicine. 2018;**3**(1):10-17

[12] Rahman HA, Abou-Elew HH, El-Shorbagy RM, Fawzy R, Youssry I. Influence of iron regulating genes mutations on iron status in Egyptian patients with sickle cell disease. Annals of Clinical and Laboratory Science. 2014;**44**(3):304-309

[13] Moreira AC, Neves JV, Silva T, Oliveira P, Gomes MS, Rodrigues PN. Hepcidin-(in) dependent mechanisms of iron metabolism regulation during infection by Listeria and Salmonella. Infection and Immunity. 2017;**85**(9): e00353-e00317

[14] Go HJ, Kim CH, Park JB, Kim TY, Lee TK, Oh HY, et al. Biochemical and molecular identification of a novel hepcidin type 2-like antimicrobial peptide in the skin mucus of the pufferfish *Takifugu pardalis*. Fish & Shellfish Immunology. 2019;**93**:683-693

[15] Chawla LS, Beers-Mulroy B, Tidmarsh GF. Therapeutic opportunities for hepcidin in acute care medicine. Critical Care Clinics. 2019;**35**(2):357-374

[16] Hanudel MR, Rappaport M, Chua K, Gabayan V, Qiao B, Jung G, et al. Levels of the erythropoietin-responsive hormone erythroferrone in mice and humans with chronic kidney disease. Haematologica. 2018;**103**(4): e141

[17] Lehtihet M, Bonde Y, Beckman L, Berinder K, Hoybye C, Rudling M, et al. Circulating hepcidin-25 is reduced by endogenous estrogen in humans. PLoS One. 2016;**11**(2):e0148802

[18] Swinkels DW, Girelli D, Laarakkers C, Kroot J, Campostrini N, Kemna EH, et al. Advances in quantitative hepcidin measurements by time-of-flight mass spectrometry. PLoS One. 2008;**3**(7):e2706

[19] Yin X, Chen N, Mu L, Bai H, Wu H, Qi W, et al. Identification and characterization of hepcidin from Nile Tilapia (*Oreochromis niloticus*) in response to bacterial infection and iron overload. Aquaculture. 2022;**546**:737317

[20] Aschemeyer S, Qiao BO, Stefanova D, Valore EV, Sek AC, Ruwe TA, et al. Structure-function analysis of ferroportin defines the binding site and an alternative mechanism of action of hepcidin. Blood. 2018;**131**(8):899-910

[21] Wrighting DM, Andrews NC. Iron homeostasis and erythropoiesis. Current Topics in Developmental Biology. 2008;**82**:141-167

[22] Corengia C, Galimberti S, Bovo G, Vergani A, Arosio C, Mariani R, et al. Iron accumulation in chronic hepatitis C: Relation of hepatic iron distribution, HFE genotype, and disease course. American Journal of Clinical Pathology. 2005;**124**(6):846-853

[23] Tangudu NK, Alan B, Vinchi F, Wörle K, Lai D, Vettorazzi S, et al. Scavenging reactive oxygen species production normalizes ferroportin expression and ameliorates cellular and systemic iron disbalances in hemolytic mouse model. Antioxidants & redox signaling. 2018;**29**(5):484-499

[24] Al-Abedy NM, Salman ED, Faraj SA. Frequency of human hemochromatosis HFE gene mutations and serum hepcidin level in iron overload β-thalassaemia Iraqi patients. Public Health. 2019; **22**(10):S275

Chapter 5

Abnormal Iron Metabolism and Its Effect on Dentistry

Chinmayee Dahihandekar and Sweta Kale Pisulkar

Abstract

Iron is a necessary micro-nutrient for proper functioning of the erythropoietic, oxidative and cellular metabolism. The iron balance in the body adversely affects the normal physiologic functioning of the body and structures in the oral cavity. Various abnormalities develop owing to improper iron metabolism in the body which reflects in the oral cavity. The toxicity of iron has to be well understood to immediately identify the hazardous effects which arise owing to it and to manage it. It has been very well mentioned in the chapter. The manifestations of defects of iron metabolism in the oral cavity should be carefully studied to improve the prognosis of the treatment of the same. Disorders related to iron metabolism should be managed for improvement in the quality of life of the patient.

Keywords: iron metabolism, anaemia, iron toxicity, manifestations in oral cavity

1. Introduction

For optimal erythropoietic function, oxidative metabolism and cellular immunity, iron is required. Cellular iron overload induces toxicity and cell death by producing free radicals and oxidising lipids, both of which are required for cellular metabolism and aerobic respiration. Due to the lack of active iron excretory mechanisms, dietary iron absorption (12 mg/day) is tightly regulated and closely balanced against iron loss. Dietary iron is found in two forms: haem (10%) and nonhaem (ionic, 90%), and both are absorbed in the apical surface of duodenal enterocytes through different mechanisms. Iron is exported via Ferroportin 1 (the only one). Absorbed iron crosses the enterocyte's basolateral membrane into the circulation (possible iron exporter), where it binds to transferrin and is transported to utilisation and storage sites transferrin-bound iron enters target cells via receptor-mediated endocytosis, mostly erythroid cells but also immune and hepatic cells. Senescent erythrocytes are phagocytosed by reticuloendothelial system macrophages; haem is metabolised by haem oxygenase, and the freed iron is stored as ferritin. Later, iron from macrophages will be exported and transferred to transferrin. The erythropoiesis demands (20e30 mg/day) need this internal iron cycle. When transferrin becomes saturated in iron-overload scenarios, excess iron is transported to the liver, the other principal storage organ for iron, creating a risk of free radical generation and tissue damage [1].

The fact that iron's redox pair (Fe(II)/Fe(III)) may have potentials varying from -300 to 700 mV, depending on the nature of the ligands and the surrounding environment, contributes to its use. Iron is abundant on the planet's surface; however, it is relatively inaccessible. This is an important aspect of iron metabolism. At neutral pH and in an oxidising environment, iron exists in the three valence state, which is seen in many common microbial environments. As a result, it is extremely difficult to dissolve. The presence of iron storage and transport proteins such as ferritin (FTN), lactoflavin (LFT) and lactoflavin (LFT) limits the amount of iron available to a microorganism residing in an animal host (LFT). Despite the fact that extremely low iron concentrations of 1 mmol (5) are usually sufficient for optimal growth yields, bacteria frequently find themselves in iron-deficient environments and must waste a significant amount of energy to acquire this metal. It is also worth mentioning that bacteria can become iron-overloaded, necessitating careful monitoring of iron intake [1, 2].

Increased iron demands, insufficient external supply and increased blood loss can contribute to iron deficiency (ID) and iron deficiency anaemia. An overabundance of hepcidin hinders iron absorption and recycling in chronic inflammation, leading to hypoferremia and iron-restricted erythropoiesis (functional iron deficiency), and finally, anaemia of chronic illness (ACD), which can advance to ACD with real ID (ACD + ID). Hereditary haemochromatosis (HH type I, caused by mutations in the HFE gene) and hereditary haemochromatosis (HH type II, caused by mutations in the hemojuvelin and hepcidin genes) can both be caused by low hepcidin expression. Changes in the transferrin receptor 2 generate HH type III, whereas mutations in the ferroportin gene induce HH type IV. All of these illnesses show signs of iron excess. In iron overload scenarios, non-transferrin bound iron develops when transferrin becomes saturated. A part of this iron (labile plasma iron) is very reactive, leading to the generation of free radicals. Free radicals induce the parenchymal cell damage associated with iron overload disorders [3].

The teeth, gingiva, oral tissues and muscles are all affected by these major metabolic anomalies of iron metabolism. These processes influencing the oral cavity must be well understood in order to block future advancement and create a comprehensive rehabilitation approach for such persons, taking into consideration the numerous consequences of improper iron metabolism [4].

2. Iron metabolism

2.1 Iron uptake

Owing to certain specific mechanisms (as explained in **Figure 1**): (1) transport mechanisms were not required for iron absorption until relatively late in evolution when the environment became oxidising and iron became insoluble, and (2) a range of sources can function as iron providers, bacterial iron assimilation happens via a variety of routes. Many bacteria, in addition, have numerous iron absorption mechanisms. This allows them to acquire iron from a variety of settings and sources. Bacteria can get iron from a number of sources, but regardless of where it comes from, it must be delivered to the cytoplasm through numerous microbial surface layers. An outside membrane, a peptidoglycan layer, and an innermost intracellular membrane are the minimum layers for Gram-negative bacteria. The periplasm, or gap between the outer and inner membranes, is where the peptidoglycan cell wall is found. Gram-positive cells, on the other hand, may only have an exterior peptidoglycan cell wall that is thick and strongly cross-linked. On the basis of the iron source

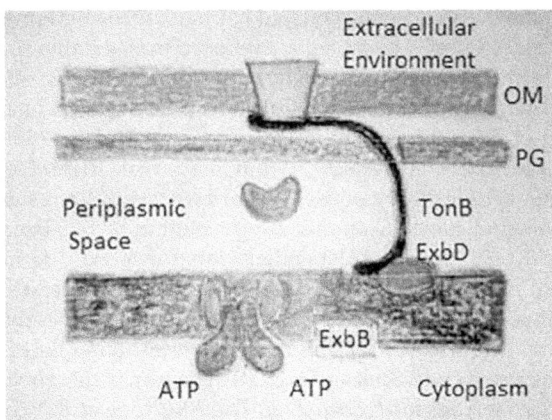

Figure 1.
Gram-negative bacteria's generalised high-affinity iron transport mechanism.) The three fundamental components are shown: (a) an outer membrane receptor protein; (b) a TonB system for activating the receptor protein; and (c) a cytoplasmic membrane-based periplasmic binding protein-dependent ABC transporter. OM stands for outer membrane; PG is for peptidoglycan; and CM stands for cytoplasmic membrane.

and the manner in which iron is mobilised, a wide range of iron transport systems may be differentiated, although they all follow a similar pattern. Passage across the outer membrane for iron complexed to a carrier requires the presence of an outer membrane receptor protein with a syntactic domain identical to that of the iron complexed to a carrier and is iron-controlled. A receptor protein is specialised for and binds to a certain iron-carrier complex, and it is occasionally generated in large quantities only when that iron complex is accessible [5].

Second, the cytoplasmic membrane proteins TonB, ExbB and ExbD are required for iron entry into the periplasm, whether it is complexed or free. Members of the ABC super transporter family are also engaged in cytoplasmic membrane transport. The transport components, in this case, include a peripheral cytoplasmic membrane, ATPase with two copies and a distinct ATP-binding site motif, as well as two hydrophobic cytoplasmic membrane proteins. In summary, an outer membrane receptor protein, a TonB system and an ABC transporter are required for iron entrance into the cytoplasm of Gram-negative bacteria. The proton-motive force and ATP, respectively, are required for passage across the outer and cytoplasmic membranes. TonB systems have broad specificity, whereas ABC transporters recognise several iron complexes if they are physically related. Outer membrane receptor proteins bind just one particular iron complex, whereas TonB systems have broad specificity [6].

The synthesis and secretion of tiny (600–1000 Da) iron-chelating molecules known as siderophores is a significant method by which bacteria acquire iron. Siderophores are made up of ordinary amino acids, nonprotein amino acids, hydroxy acids, and their production does not need ribosomes despite the presence of amide bonds. Instead, a thiotemplate technique is used, which is quite similar to the one used to make some peptide antibiotics. The mechanism of iron release from siderophores is unknown. Free siderophores, or modified forms, are discharged into the medium when ferrisiderophores enter the cytoplasm. Enzymatic reduction of iron is considered to be the release mechanism since siderophores: (1) bind $Fe(II)$ less readily than $Fe(III)$; and (2) the cytoplasm is a reducing environment [7].

Microbes that can live in oxygen-depleted habitats, such as swamps, intestines and marshes, or acidic environments, where reduced iron is stable and soluble, benefit from the ferrous iron transfer. Fe(II) may enter the periplasm through holes in the outer membrane, and bacteria can transport it through the inner membrane through a number of mechanisms. Some of these cytoplasmic membrane transporters have a broad transition metal selectivity but just a weak affinity for ferrous iron. There are, however, systems that exclusively work with Fe(II) as a substrate. The feo operon encodes one such mechanism that is important in certain bacteria (feoABC) [8]. Members of the OFeT (oxidase-dependent iron transporter) family, which were initially discovered in lower eukaryotes, are another widely dispersed group of Fe(II) transporter proteins. Finally, certain aerotolerant bacteria, such as the Gram-positive *Streptococcus mutans*, acquire iron by converting surface-bound Fe(III) to Fe via a reductase that is exposed on the cell surface (II). The iron is subsequently delivered to the cytoplasm by a ferrous ion transporter. The ABC type of ferric iron acquisition mechanism is found in a variety of Gram-negative taxa, including Serratia, where it was identified and named Sfu type transport. Fe(III) is accepted by a periplasmic binding protein, which then delivers it to the transporter's cytoplasmic components, which internalise the iron. Uptake systems with outer-membrane components can also work in tandem with ferric iron transporters [9].

The bulk of iron in animals is found intracellularly in the form of heme (Hm). Hm, in turn, is a prosthetic group of proteins that includes haemoglobin (Hb), myoglobin and Hm-containing proteins like cytochromes. Iron assimilation routes that detect free Hm are similar to those that identify iron–siderophore complexes; they need (1) a TonB-dependent outer-membrane receptor protein; and (2) an ABC transporter for cytoplasmic membrane crossing [10]. Hm can be removed from Hb by a variety of genera. TonB-dependent Hb-binding proteins are found in the outer membranes of Neisseria and Haemophilus spp. Surprisingly, both of these taxa contain additional TonB-dependent receptors that let them get iron from Hb–Hp complexes. These Hb–Hp receptors might be made up of two distinct proteins. *Serratia marcescens* has a unique mechanism for the first steps in getting iron from Hm or Hb. This bacterium secretes a tiny protein (HasA) that acts as a hemophore via an ABC transporter (Hbp). Only Haemophilus strains have been shown to use Hm in conjunction with hemopexin. The mechanism is not fully understood, but it appears that three genes are necessary, one of which appears to encode a big secreted Hbp (HxuA). HxuA binds hemopexin, removes it, and transports it to an outer-membrane receptor [11].

Iron trafficking exemplifies the cycle economy. During erythrocyte phagocytosis, the majority of iron (20–25 mg/day) is recycled by macrophages; only 1–2 mg of iron is absorbed daily in the stomach, compensating for a loss of the same amount (**Figure 2**) [13]. The duodenum is the location of controlled non-heme iron uptake; nonheme iron is imported from the lumen via the apical divalent metal transporter 1 after duodenal cytochrome B reductase converts ferric to ferrous iron (DCYTBH) (DMT1). There are no known mechanisms by which heme iron absorbs more than non-heme iron. Non-utilised iron in enterocytes is either retained in ferritin (and lost by mucosal shedding) or exported to plasma through basolateral membrane ferroportin (and lost with mucosal shedding) [14].

Iron availability influences the expression of genes that code for proteins required for high-affinity iron absorption. Fur is a crucial regulatory protein found in most Gram-negative and Gram-positive bacteria with low GC content DNA. Fur is an Apo-repressor, a short histidine-rich polypeptide that binds DNA in the presence of its corepressor Fe(II). Fur's negative regulation of genes does not fully explain iron's regulatory actions. Although Fur represses most iron-regulated genes under iron-rich environments, some are positively controlled by Fur, and others are only activated by iron in the absence of Fur (**Figure 3**) [15].

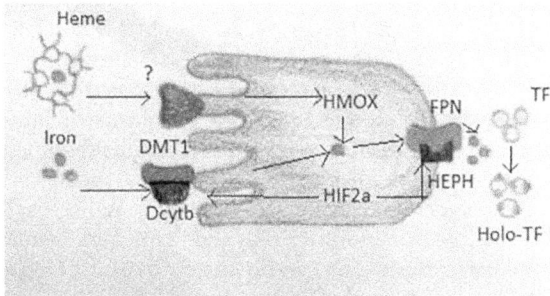

Figure 2.
On the luminal side of the enterocyte, the metal transporter DMT1 takes up ferrous iron that has been reduced by DCYTB. After ferrous iron is oxidised to ferric iron by hephaestin, iron not utilised inside the cell is either stored in ferritin (FT) or exported to circulating transferrin (TF) by ferroportin (FPN) (HEPH). Local hypoxia stabilises hypoxia-inducible factor (HIF)-2, which promotes the expression of the apical (DMT1) and basolateral (FPN) transporters. Heme is transformed to iron by heme oxygenase once it enters the cell by an unknown process [12].

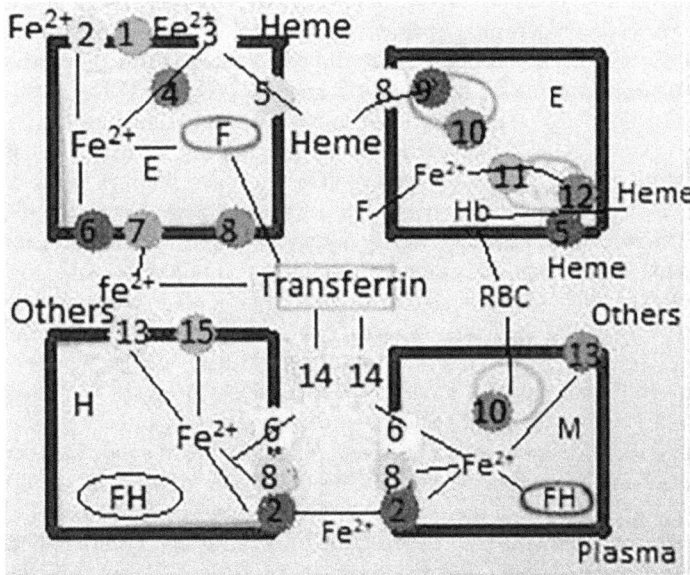

Figure 3.
Main iron metabolism routes in animals (based on Munoz et al.2). Key: 1, ferrireductase; 2, divalent metal transporter (DMT1); 3, haem protein carrier 1 (HPC1); 4, haem oxygenase; 5, haem exporter; 6, ferroportin (Ireg-1); 7, hephaestin/caeruloplasmin; 8, transferrin receptor-1 (TfR1); 9, transferrin receptor-1 (TfR1) complex; 10, natural resistance macrophage protein-1 (Nramp-1); 11, mitoferrin; 12, mitochondrial haem exporter (Abcb6); 13, others: bacteria, lactoferrin, haemoglobinehaptoglobin, haemehaemopexin, and so on; 14, caeruloplasmin; 15, transferrin receptor-2 (TfR2).

2.2 Iron distribution

Transferrin binds to iron in the bloodstream and distributes it to storage and use sites. Only 30–40% of transferrin's iron-binding capacity is used in ordinary physiological circumstances; hence, transferrin-bound iron is only w4 mg, yet it is the most significant dynamic iron pool. Transferrin-bound iron penetrates target cells, predominantly erythroid cells, but also immune and hepatic cells, via a highly specialised method of receptor-mediated endocytosis (**Figure 1**). Patches of cell-surface membrane bearing receptor–ligand complexes invaginate to create clathrin-coated endosomes as distinct transferrin binds to transferrin receptor 1 (TfR1) at the plasma membrane (siderosomes) [16]. A ferrireductase reduces $Fe3+$ to $Fe2+$, which is subsequently transferred to the cytoplasm by DMT1, while TfR1 is recycled to the cell membrane and transferrin is lost. Mitoferrin, a mitochondrial iron importer, is important in providing iron to ferrochelatase for insertion into protoporphyrin IX and to produce haem (the penultimate step of mitochondrial haem production) within the erythroblast (**Figure 1**). There are some indications that iron might be transported straight from the siderosomes to the mitochondria in growing erythroid cells. Finally, haem exporters transport haem from mitochondria to cytosol and eliminate excess haem from erythroid cells (**Figure 1**) [16].

2.3 Iron storage

As senescent erythrocytes are phagocytosed by RES macrophages, haemoglobin iron turnover is high. Haem is metabolised by haem oxygenase within the phagocytic vesicles, and the liberated $Fe2+$ is transported to the cytoplasm by NRAMP1 (natural resistance-associated macrophage protein-1), a transport protein related to DMT1 (**Figure 1**). Macrophages may also acquire iron from bacteria and apoptotic cells, as well as from plasma via the actions of DMT1 and TfR1 (**Figure 1**) [17]. Iron may be stored in the cells in two ways: ferritin in the cytosol and haemosiderin in the lysosomes when ferritin is broken down. Haemosiderin is found in just a small percentage of normal human iron reserves, primarily in macrophages, but it rises substantially when the body is overloaded with iron. Iron storage in macrophages is also safe since it does not cause oxidative damage. Ferroportin 1, the same iron-export protein found in the duodenal enterocyte, and caeruloplasmin2 are largely responsible for iron export from macrophages to transferrin (**Figure 1**) [18]. Macrophage iron recycling provides the majority of the iron necessary for the daily synthesis of 300 billion red blood cells (20–30 mg). While a result, internal iron turnover is required to satisfy the bone marrow needs for erythropoiesis, as daily absorption (1–2 mg) only balances daily loss. 1–3 The liver is the other major iron storage organ, and the production of free radicals and lipid peroxidation products in iron-overload conditions can lead to hepatic tissue damage, cirrhosis, and hepatocellular cancer [19]. TfR1 and TfR2 mediate the liver's absorption of transferrin-bound iron from plasma (**Figure 1**), however, it can also get iron from non-transferrin-bound iron (through a carrier-mediated mechanism similar to DMT1), ferritin, haemoglobine–haptoglobin complexes, and haeme–haemopexin complexes. Ferroportin 1 is thought to be the sole protein that mediates the export of iron from hepatocytes, which is then oxidised by caeruloplasmin and attached to transferrin2 (**Figure 1**). Heart failure is the primary cause of death in individuals with untreated hereditary haemochromatosis or transfusion-associated iron overload, thus iron storage in cardiomyocytes is of significant interest. Excess iron in cardiac cells can cause oxidative stress and impair myocardial function owing to DNA damage caused by hydrogen peroxide via the Fenton reaction [20].

2.4 Regulation of iron homoeostasis

Body iron reserves, hypoxia, inflammation and erythropoiesis rate all influence iron absorption by duodenal enterocytes. The crypt programming model and the hepcidin model are two regulatory models that have been presented as potential contributors to iron absorption control [21].

Enterocytes in the crypts of the duodenum take up iron from the plasma via TfR1 and TfR2, according to the crypt programming hypothesis. The interaction of cytosolic iron regulatory proteins (IRPs) 1 and 2 with iron-responsive elements is controlled by intracellular iron content (IREs). IRP1 binds to the IREs of TfR1, DMT1, and ferroportin 1 mRNA in the absence of iron, stabilising the transcript, allowing translation to occur and the proteins to be synthesised. As a result, increased IRP-binding activity indicates low body iron reserves, which leads to overexpression of these proteins in the duodenum, boosting dietary iron absorption. When IRPs attach to ferritin mRNA's IREs, the transcript's translation is interrupted and synthesis is halted. As a result, ferritin concentrations are inversely controlled, increasing in iron-rich states and decreasing in iron-deficient conditions [22].

The hepcidin model proposes that hepcidin is produced mainly by hepatocytes in response to the iron content of the blood. Then, hepcidin is secreted into the bloodstream and interacts with villous enterocytes to regulate the rate of iron absorption by controlling the expression of ferroportin 1 at their basolateral membranes. The binding of hepcidin to ferroportin 1 initially causes Janus kinase 2-mediated tyrosine phosphorylation of the cytosolic loop of the carrier protein, phosphorylated ferroportin 1 is then internalised, dephosphorylated, ubiquitinated and ultimately degraded in the late endosome/lysosome compartment. Ferroportin 1 molecules, present in macrophages and liver, also targets for hepcidin [23].

The sensing process most likely includes local iron-induced synthesis of bone morphogenic proteins (BMPs) such as BMP6 within normal iron concentration limits. BMP6 interacts with hepatocyte cell surface BMP receptors (BMPRs) I and II, as well as the BMP coreceptor, haemojuvelin (HJV), triggering an intracellular signal by phosphorylation of small mothers against decapentaplegic (Smad) proteins. Before translocating to the nucleus and triggering hepcidin expression14, phosphorylated Smad1, Smad5 and Smad8 form a complex with the shared mediator Smad4 (**Figure 2**). The soluble form of HJV (sHJV), whose release (HJV shedding) is prevented by rising extracellular iron concentrations, is thought to compete with its membrane-anchored counterpart for BMPR binding, resulting in iron-sensitive hepcidin expression16 (**Figure 2**). Other mediators and modulators, including Smad6 and Smad7, may be stimulated by iron, and these mediators and modulators appear to dampen the signal for hepcidin activation (**Figure 2**) [24].

2.5 Effects of inflammation on iron homoeostasis and erythropoiesis

Cancer, rheumatoid arthritis, inflammatory bowel disease, congestive heart failure, sepsis and chronic renal failure are all known to induce persistent inflammation. This anaemia might be caused by the underlying process activating the immune system, as well as immunological and inflammatory cytokines such as tumour necrosis factor alpha (TNFa), interferon-gamma (IFNg), interleukins (IL) 1, 6, 8, and 10. Several pathophysiological processes (cytokines) may be implicated in anaemia of chronic disease (ACD) (**Figure 3**) [25]:

1. Dyserythropoiesis, red blood cell destruction, and increased erythrophagocytosis cause a reduction in red blood cell half-life (TNFa).

2. Inadequate EPO responses for the degree of anaemia in most, but not all, patients, such as those with juvenile chronic arthritis with systemic start (IL-1 and TNFa).

3. Erythroid cell response to EPO is impaired (IFNg, IL-1, TNFa, hepcidin).

4. Erythroid cell growth and differentiation are slowed (IFNg, IL-1, TNFa, and a1-antitrypsin).

5. Pathological iron homoeostasis caused by increased DMT1 (IFNg) and TfR (IL-10) expression in macrophages, decreased ferroportin 1 expression in enterocytes (inhibition of iron absorption) and macrophages (inhibition of iron recirculation), and increased ferritin production '(TNFa, IL-1, IL-6, IL-10) (increased iron storage) Inflammatory cytokines like IL-6' activate Janus kinases, which phosphorylate Stat3 and activate it, which upregulate hepcidin transcription. Stat3 translocation to the nucleus and binding to the Stat response element in the proximal promoter of the hepcidin gene leads to enhanced hepcidin release. This element appears to be controlled by Smad activation, which is necessary for complete promoter activity, via the adjacent BMP-responsive element. The SmadeStat complex, which puts the distal and proximal areas of the hepcidin promoter into physical contact, is hypothesised to interact with a distal BMP responsive element location. As a result, it appears that Smad signalling is critical for the appropriate staging of the inflammatory response. Stat3 activation has also been demonstrated to modulate hepcidin levels without producing inflammation (for example, people with glycogen storage disease type 1a who had hepatic adenomas overexpressed hepcidin due to Stat3 activation)46 (**Figure 2**). Stress mechanisms signalling through the cellular endoplasmic reticulum unfolded protein response have also been shown to stimulate hepcidin production. The hepatic acute-phase response to LPS, IL-6 and IL-1b has been related to the unfolded protein response, suggesting that hepcidin gene expression may be regulated by another layer of endogenous regulation during inflammation (**Figure 2**). Low blood iron and reduced transferrin saturation are produced by iron diversion to the RES (functional iron deficit, FID), iron-restricted erythropoiesis, and mild-to-moderate anaemia, despite normal or high serum ferritin levels [26].

3. Defects of iron metabolism

3.1 Iron deficiency

In the human body, there is a balance between iron absorption, iron transit and iron storage under physiological circumstances. ID and iron deficiency anaemia (IDA) can be caused by a combination of three risk factors: higher iron needs, restricted external supply and increased blood loss [27]. There are two types of ID: absolute and functional. Iron reserves are reduced in absolute ID; in functional iron deficiency (FID), iron stores are full but cannot be mobilised as quickly as needed from the RES macrophages to the bone marrow. Diagnostic tests with values are given in **Table 1**.

Laboratory tests	Conventional units
Serum Iron	50–150 ug/dl
Transferrin	200–360 mg/dl
Transferrin Saturation	20–50%
Ferritin (Ft)	30–300 ng/ml
Soluble transferrin receptors	0.76–1.76 mg/l
Ratio of sTfR to Serum Ft	<1
Haemoglobin	12–16 g/dl (women); 13–17 g/dl (men)
MCV	80–100 fl
Red Cell Distribution	11–15
MCH	28–35 pg
Hypochromic red cells	<5%
Reticulocyte haemoglobin content	28–35 pg

Table 1.
Depicting tests required for determination of iron metabolism anaemia.

3.1.1 Iron deficiency anaemia

Patients with low Hb (13 g/dl for males and 12 g/dl for women), TSAT (20%) and ferritin (30 ng/ml) concentrations but no indications of inflammation should be evaluated to have IDA. Instead of 'mean corpuscular volume (MCV)', the MCH has emerged as the most significant marker for red cells for identifying ID in RBCs, which are circulating (**Figure 1**). MCV is a generally available and reliable measurement, although it is a late indication in individuals who are not bleeding actively [28]. When MCV is low, thalassaemia must be considered a differential diagnosis. When there is a concurrent folate deficiency or vitamin B12, reticulocytosis post-bleeding, early response to oral iron therapy, alcohol use, or moderate myelodysplasia, individuals may present with IDA but no microcytosis. Human serum contains a shortened, 'soluble version of the transferrin receptor (sTfR)', whose concentration is proportional to the total number of cell surface transferrin receptors [29]. Although the amount is not defined and depends on which reagent kit is used, normal median values are 1.2–3.0 mg/l. Even during chronic illness anaemia, increased sTfR values suggest ID. Elevated erythropoietic activity without ID, during reticulocytic crises, and in congenital dyserythropoietic anaemias are all examples of increased sTfR levels. Lower sTfR levels, on the other hand, might indicate a reduction in the number of erythroid progenitors. Despite the fact that sTfR levels in simple IDA are generally high or extremely high, they are not usually necessary for diagnosis [30].

3.1.2 Anaemia of chronic disease

The following should be present in patients with chronic disease anaemia (ACD), also known as anaemia of inflammation: Hb concentration of 13 g/dl for men and 12 g/dl for women; a low TSAT (20%) but normal or increased serum ferritin concentration (>100 ng/ml) or low serum ferritin concentration (30e100 ng/ml) Evidence of chronic inflammation (e.g. elevated CRP); and a s ACD, like FID, is common in people with inflammatory illness but no visible blood loss (e.g. rheumatoid arthritis, renal failure or chronic hepatitis) [31].

3.2 Iron overload

Levels of Hepcidin are excessively lower-degree of overload of iron in idiopathic iron overload illness and primary haemochromatosis. This is due to mutations in the genes that code for 'HFE (haemochromatosis type 1)', 'haemojuvelin (HJV; juvenile haemochromatosis 2a)', and 'transferrin receptor 2 (TfR2; haemochromatosis type 3)'; these mutations cause hepcidin synthesis to be dysregulated [32]. The only exceptions are mutations that disrupt hepcidin or ferroportin (juvenile haemochromatosis 2b) (haemochromatosis type 4). Low plasma hepcidin causes high ferroportin levels, allowing for greater iron absorption, hepatic iron overload and low iron levels in macrophages. In addition, non-transferrin bound iron emerges as transferrin gets saturated in iron-overload situations. A portion of this labile plasma iron is extremely reactive, resulting in the production of free radicals. Despite the fact that the HFE gene has at least 32 mutations, the most prevalent form of haemochromatosis type 1 is caused by the missense Cys282Tyr mutation. Haemochromatosis type 1 is a disease with a wide range of penetrance and heterogeneity, although the Cys282Tyr mutation is found in the great majority of people with the disorder. Because the Cys282Tyr mutant HFE protein is unable to bind b2 microglobulin, it does not reach the cell membrane, resulting in a misfolded, non-functional protein. Iron overload can be caused by mutations in the ferroportin gene (haemochromatosis type 4) that result in the loss of iron-export capacity, hyperferritinaemia with no increase in transferrin saturation, and macrophage iron overload, or a loss of hepcidin-binding activity, which has been linked to iron overload. Plasma hepcidin levels rise in cases of secondary iron overload-induced by persistent transfusion treatment (e.g. severe thalassemia, aplastic anaemia, etc.), prompting ferroportin breakdown. Increased amounts of diferric transferrin, which are elevated in iron overload, promote TfR2 expression at the hepatocyte membrane. When diferric transferrin binds to TfR2, HJV cleavage by furin is blocked, inhibiting the release of soluble HJV and resulting in enhanced cell-surface HJV-mediated response to BMPs and higher hepcidin levels. Iron absorption from the stomach is restricted, macrophage export is inhibited, and iron storage is increased when ferroportin levels are low [33].

3.3 Assessment of defective iron metabolism

3.3.1 Laboratory assessment of ID

Measurements indicating iron depletion in the body and measurements indicating iron-deficient red cell production are the two types of laboratory tests used to investigate ID (**Table 1**). The right mix of these blood tests will aid in determining the precise diagnosis of anaemia and ID status (**Figure 1**).

3.3.2 Assessment of iron overload

The first step in diagnosing iron overload is to suspect it (e.g. dark skin, fatigue, arthralgia, cardiomyopathy, hepatomegaly, endocrine disorder, etc). However, aberrant TSAT (>45 per cent) and/or elevated ferritin in serum (>200 ng/ml in women, >300 ng/ml in males) are commonly discovered. In practice, normal transferrin saturation can be used to rule out the possibility of iron overload. The sole exception is the occurrence of an inflammatory state, which might disguise an increase in TSAT, which is why CRP and transferrin saturation should be checked jointly.

In non-iron-overload circumstances, such as significant cytolysis (eg. acute hepatitis), which raises plasma serum iron and/or hepatic failure, reduces plasma transferrin concentrations, elevated TSAT can be detected. Other causes of hyperferritinaemia should be checked out in the presence of elevated ferritin in serum but not increased TSAT (eg. cell necrosis, alcohol, inflammation, metabolic disorder, etc). The clinical context, as well as testing Hb (to rule out chronic inflammatory anaemia), transaminases, cancer and prothrombin index, can readily remove any difficulties in interpreting TSAT readings (to exclude hepatic disease) [34].

The second diagnostic step, particularly in Caucasian individuals, is to rule out HFE mutations in gene. Because further mutations in HFE are exceedingly rare, the HFE genotype is frequently regarded as 'wild type' in clinical practice, once the presence of the two most prevalent (Cys282Tyr and His63Gly) mutations has been ruled out. Nonetheless, the potential of a family problem should be addressed at all times: a dominant disorder is usually indicative of ferroportin disease [35].

Before beginning costly and time-consuming searches for mutations in additional genes, the third diagnostic step is to establish increased total body iron. The exact molecular diagnosis, which needs evidence of the nucleotide mutation at the DNA level, is the fourth stage. However, the efficacy of molecular diagnostics is frequently questioned because it is costly, time-demanding and, in certain situations, unable to produce a precise diagnosis [36].

4. Iron metabolism and the oral health

4.1 Iron deficiency anaemia

The most prevalent kind of anaemia is iron deficiency anaemia (IDA), which affects more women than males. Due to persistent blood loss associated with heavy menstrual flow, it is estimated that 20% of women of reproductive age in the United States are iron deficient. Furthermore, 2% of adult males are iron deficient due to persistent blood loss caused by gastrointestinal illnesses including peptic ulcer, diverticulosis, or cancer [37].

4.1.1 Symptoms

Atrophic glossitis (AG), extensive oral mucosal atrophy and pain or burning feeling of the oral mucosa are some of the oral symptoms and indicators. However, it is yet unknown if IDA patients may experience distinct oral signs and, if so, what percentage of IDA patients experience these oral manifestations. Burning sensation of the oral mucosa (76.0 per cent), lingual varicosity (56.0 per cent), dry mouth (49.3%), OLP (33.3 per cent), AG (26.7 per cent), RAU (25.3 per cent), numbness of the oral mucosa (21.3 per cent) and taste dysfunction (12.0 per cent) were the most commonly manifested oral manifestations. IDA patients had considerably greater rates of all oral symptoms, such as oral mucosa burning, lingual varicosity, dry mouth, oral mucosa numbness, and taste impairment than healthy controls.

4.1.2 Pathophysiology

Anaemia sufferers have low haemoglobin levels, which means they do not get enough oxygen to their mouth mucosa, causing it to atrophy. Iron deficiency can induce oral mucosa atrophy because iron is required for proper oral epithelial cell activity, and in an iron deficiency condition, oral epithelial cells turn over more quickly, resulting in an atrophic or immature mucosa. The health of the oral epithelium is linked to iron and vitamin B12.

In BMS patients, long-term dry mouth and iron or vitamin B12 deficiency may produce at least partial atrophy of the tongue epithelium, however, the change is so mild that clinical visual examination cannot detect it. As a result, spicy chemicals in saliva might readily permeate past the atrophic epithelium into the subepithelial connective tissue of the tongue mucosa, irritate free sensory nerve endings, and cause tongue burning and numbness. A minor sign of BMS was loss or malfunction of taste. Because the taste cells in taste buds can only sense dissolved compounds, the chemical components should be dissolved in saliva.

The majority of BMS patients were found to have xerostomia. In BMS patients, decreased saliva output leads to a loss or malfunction of taste. Oral candidiasis, vitamin B12 insufficiency, iron deficiency and medicine have all been linked to taste loss or malfunction. Femiano et al. have looked into the causes of taste disturbance in BMS patients. Of the 142 BMS patients, 61 had a documented history of drug use that interfered with taste perception, 35 had pathologies or a past history of drug use that were known to impact the gustatory system, and the other 46 had no related disease or regular drug use [38].

Varicosities are abnormally dilated, and convoluted veins are observed on the ventral surface of the tongue in elderly people due to a decrease in connective tissue tone that supports the veins. Furthermore, xerostomia is a prevalent issue that affects 25% of the elderly population. Xerostomia can be caused by a variety of developmental, iatrogenic, systemic and local causes. Older individuals, on the other hand, are more likely to develop xerostomia, as a result of pharmaceuticals, as they are more likely to use drugs that induce xerostomia to treat their systemic or psychotic diseases. The average age of 399 BMS patients in Wang Y et al's research was 59.7 years. As a result, it is not unexpected that 92.5 per cent of 399 BMS patients had lingual varicosity and 75.7 per cent had dry mouth. Oral candidiasis is more common in persons with xerostomia because normal and adequate saliva can offer cleaning and antibacterial action. We believe that the candidiasis on the tongue surfaces of BMS patients is attributable, at least in part, to the high prevalence of dry mouth (75.7%).

4.1.3 Management

4.1.3.1 Oral iron

In most therapeutic situations, oral iron supplementation is sufficient. In the absence of inflammation or severe continuous blood loss, oral iron, usually in the form of ferrous salts, can be used to treat anaemia if large dosages are tolerated. Although traditional knowledge holds that up to 200 mg of elemental iron per day is necessary to treat IDA, this is erroneous and lesser amounts can be effective as well.

Early research suggested that taking iron with vitamin C might help with iron absorption because more ferrous iron is kept in the solution. However, findings suggest that co-administration of these drugs might cause serious toxicity in the gastrointestinal tract. Furthermore, while taking oral iron away from meals is often suggested to increase absorption, it also increases gastric intolerance, which reduces compliance. Furthermore, some antibiotics (primarily quinolones, doxycycline, tetracycline, chloramphenicol, or penicillamine), proton pump inhibitors, and anti-acid medication (aluminium, bicarbonate, zinc, or magnesium salts), levodopa, levothy-roxine, cholestyramine, phytates (high-fibre diets), soy products, ibandronate, etc.

Non-absorbed iron salts, on the other hand, can produce a variety of highly reactive oxygen species, such as hypochlorous acid, superoxides and peroxides, which can cause digestive intolerance, resulting in nausea, flatulence, abdominal pain, diarrhoea or constipation and black or tarry stools, as well as relapsed inflammatory bowel disease. As a result, smaller iron salt dosages (e.g. 50–100 mg elemental iron) should be advised. The Ganzoni method may be used to determine the total iron deficiency (TID): TID (mg) 14 weight (kg) 3 (ideal Hb e actual Hb) (g/dl) 3 0.24 + depot iron (500 mg). An individual, weighing 70 kg, with a haemoglobin level of 9 g/dl would have a body iron shortfall of around 1400 mg, according to this calculation.

4.1.3.2 Parenteral iron

Parenteral iron is traditionally used to treat intolerance, contraindications, or an insufficient response to oral iron. However, in circumstances when there is a limited time until surgery, severe anaemia, especially if it is accompanied by considerable continuous bleeding or the use of erythropoiesis-stimulating drugs, parenteral iron is now an effective therapy. Because they provide various benefits over oral supplements, modern intravenous iron formulations have emerged as safe and effective options for anaemia therapy. In normal persons, intravenous iron delivery allows for a fivefold erythropoietic response to substantial blood loss anaemia,19 Hb begins to rise after a few days, the percentage of responsive patients increases, and iron reserves are replenished. Increasing iron reserves is beneficial, especially for patients using erythropoiesis-stimulating drugs. In clinical practice, iron gluconate, iron sucrose, high molecular weight iron dextran (HMWID), low molecular weight iron dextran (LMWID), ferric carboxymaltose, iron isomaltoside 1000 and Ferumoxytol are the most commonly used products.

4.1.4 Changing microflora in patients with ida and its corelation with infective endocarditis

The link between oral microbiota and IE (infectious endocarditis) has long been known. Infectious endocarditis is caused by opportunistic infections in normal oral flora entering the circulation through everyday mouth washing or invasive dental treatments. In vitro iron deficiency causes a dramatic change in the oral microbiota community, with higher proportions of taxa linked to infective endocarditis. Iron deficiency anaemia is utilised as an in vivo model to evaluate the association between insufficient iron availability, oral microbiota, and the risk of IE, as well as to perform population amplification research. In a research by Xi R et al., 24 patients with primary iron deficiency anaemia (IDA) from the haematology department of West China Hospital, Sichuan University, and 24 healthy controls were included from 2015.6 to 2016.6. The dental plaque microbiota of 24 IDA (iron-deficiency anaemia) patients and 24 healthy controls were compared using high-throughput sequencing. Internal diversity in the oral flora is reduced as a result of iron shortage. Corynebacterium, Neisseria, Cardiobacterium, Capnocytophaga and Aggregatibacter had considerably greater proportions in controls, whereas Lactococcus, Enterococcus, Lactobacillus, Pseudomonas and Moraxella had significantly larger proportions in the IDA group (P 0.05). Lactococcus, Enterococcus, Pseudomonas and Moraxella relative abundances were substantially inversely linked with serum ferritin concentrations (P 0.05). In vivo iron shortage altered the organisation of the oral microbiome population. When compared to healthy controls, people with IDA had lower total bacterial diversity and different taxonomic makeup. The IDA group had greater proportions of the genera Lactococcus, Enterococcus, Pseudomonas and Moraxella, whose abundance was likewise statistically and adversely linked with serum ferritin levels. Because the IDA group has a high rate of penicillin resistance, the typical use of preventive penicillin may be ineffective. The findings of a disproportionate oral microbiota suggest that more targeted antibiotic usage with various groups may be required before dangerous oral surgeries.

4.2 Iron overload

Hemochromatosis is the abnormal accumulation of iron in parenchymal organs, leading to organ toxicity. It is the most common inherited liver disease in whites and the most common autosomal recessive genetic disorder. Genetic haemochromatosis (GH), which is related to the HFE gene p.Cys282Tyr mutation, is the most common form of inherited iron overload disease in European population descendants.

4.2.1 Symptoms

The classic tetrad of manifestations resulting from hemochromatosis consists of: (1) cirrhosis, (2) diabetes mellitus, (3) hyperpigmentation of the skin and teeth, and (4) cardiac failure. Clinical consequences also include hepatocellular carcinoma, impotence and arthritis (**Figures 4** and **5**) [9].

Symptoms can vary from burning mouth syndrome to bald and inflamed tongue [9].

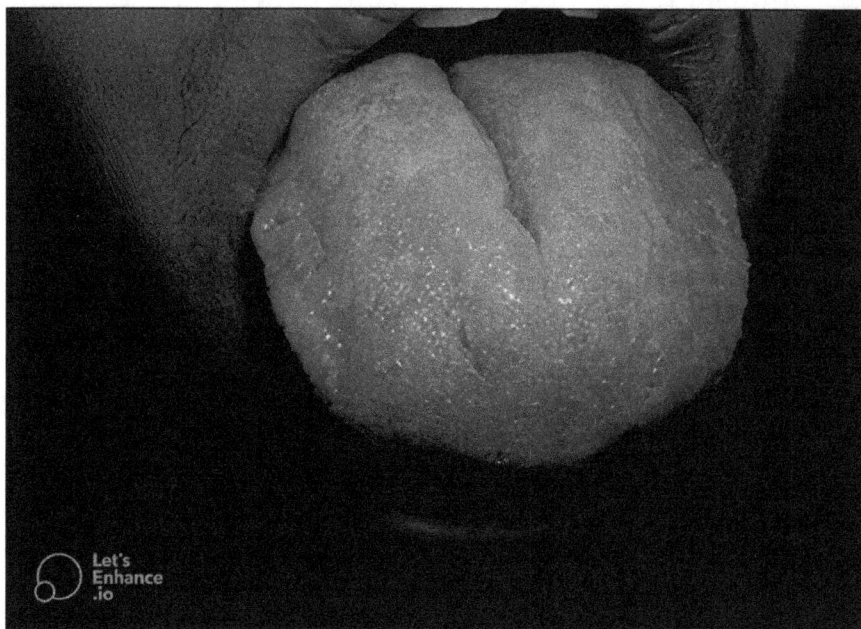

Figure 4.
Tongue anomaly of iron deficiency anaemia.

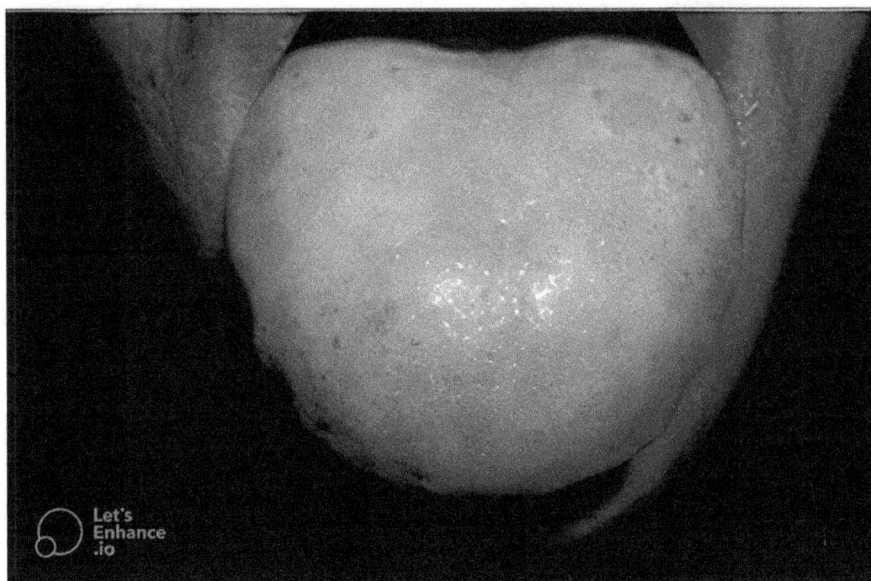

Figure 5.
Balding of tongue seen due to iron deficiency anaemia.

4.2.2 Pathophysiology

Periodontitis is linked to an inflammatory response triggered by changes in the subgingival biofilm. Inflammation causes iron sequestration inside macrophages in healthy people, depriving bacteria of iron. Iron bioavailability in biological fluids, particularly those of the oral cavity, is enhanced in GH patients with excessively high TSAT, resulting in an increased risk of severe periodontitis. The existence of iron deposits in oral tissues of haemochromatosis patients has also been documented in the literature (**Figure 6**). The majority of people with haemochromatosis are now asymptomatic, and the skin and mucosal colouration caused by iron deposits have improved dramatically. The occurrence of asymptomatic iron deposits in oral tissues, however, cannot be ruled out [10, 11].

Iron is connected with transferrin in plasma, which increases its bioavailability for cells. The ratio between the total number of iron-binding sites on patient plasma transferrin and the number of binding sites occupied by iron is known as transferrin saturation (TSAT). TSAT is normally seen in the range of 20% to 45 per cent. Hepcidin regulates systemic iron metabolism, and its expression level is tuned to TSAT to regulate plasma iron levels. Hepcidin insufficiency is a symptom of GH, which is caused by a change in the HFE-linked transduction signalling pathway. TSAT levels rise as a result of the iron outflow from macrophages and enterocytes. Non-transferrin-bound iron (NTBI), an aberrant biochemical type of iron, arises in the plasma when TSAT surpasses 45 per cent. The liver and heart are particularly vulnerable to NTBI, which explains why the typical type of GH causes hepatic cirrhosis and diabetes. However, in the absence of cirrhosis or diabetes, the majority of GH patients remain asymptomatic or have chronic tiredness, abnormal serum transaminase levels, rheumatism, and osteoporosis. Cells manufacture ferritin to store excess

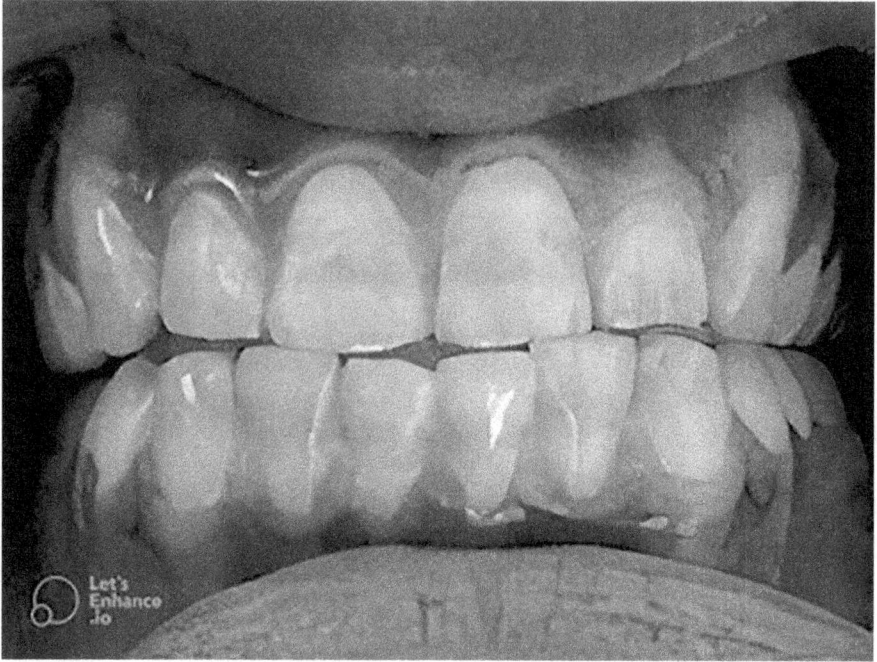

Figure 6.
Staining of teeth seen due to iron deficiency anaemia.

iron in order to avoid iron toxicity. As a result, the tissue iron reserves are reflected in plasma ferritin levels. The standard treatment is phlebotomy therapy, which is used to take out excess iron and then prevent it from being reconstituted. The gold standard for both initial treatment and maintenance therapy, according to the leading international standards, is serum ferritin levels of less than 50 g/L [13].

4.2.3 Management

Iron depletion would lessen or eliminate the risk of iron-mediated tissue harm, according to the earlier reasoning for blood removal in all patients with haemochromatosis. This may help to avoid or lessen the severity of some haemochromatosis problems after iron deficiency. Dyspnoea, pigmentation, weariness, arthralgia, or hepatomegaly may be reduced, and diabetes mellitus management and left ventricular diastolic function may be improved. The progression of hepatic cirrhosis, as well as the increased risk of primary liver cancer, hyperthyroidism and hypothyroidism, are largely unaffected.

Standard therapy for most patients with haemochromatosis and iron overload is weekly blood removal to bring ferritin levels into the low reference range (20–50 ng/ml), followed by a life-long maintenance phlebotomy schedule to maintain ferritin levels at around 50 ng/ml, for preventing or treating iron overload. The number of units to be removed can be calculated using the following formula: 1 ng/ml ferritin corresponds to nearly 8 mg mobilisable iron in the absence of hepatic necrosis or another source of inflammation that causes hyperferritinaemia, and a 500 ml blood unit contains approximately 200 mg iron. To achieve iron depletion, a patient with serum ferritin of 1000 ng/ml will likely require the removal of 40 units of blood. Traditional phlebotomy or erythrocytapheresis can be used to remove blood. Traditional phlebotomy (250–500 mL once or twice weekly during the initial phase, depending on patient's characteristics and level of iron overload, followed by 500 mL every 2–4 months for the rest of one's life) is effective for iron depletion, but it necessitates normal erythropoiesis and frequent visits to a healthcare facility, and some patients report intolerance. Blood taken for therapeutic phlebotomy at blood donation facilities can be used to supplement the blood supply for transfusion, according to new US Food and Drug Administration rules (Title 21, Code of Federal Regulations, Section 640.120). (21 CFR 640.120). Isovolaemic, large-volume erythrocytapheresis, on the other hand, removes more blood erythrocytes each session than phlebotomy while leaving plasma proteins, coagulation factors, and platelets alone. As a result, therapeutic erythrocytapheresis is a quick and safe procedure that may be recommended in the early stages of treatment for individuals with significant iron excess. Although a single therapeutic erythrocytapheresis session is more expensive, the overall expenditures to cause iron depletion are comparable to or less expensive than therapeutic phlebotomy; yet, the treatment is only available in limited quantities (special apparatus and facilities, trained personnel, etc). Both treatments, however, have comparable side effects: transitory hypovolaemia; weariness (Hb levels should not go below 11 g/dl); enhanced iron absorption; citrate response (erythrocytapheresis alone); or iron insufficiency if proper monitoring is not performed. Iron chelation therapy, on the other hand, is seldom optimal for patients with haemochromatosis, unless they are unable to undertake phlebotomy therapy due to expense, probable toxicity and a lack of proof of benefits. Finally, while dietary restrictions (e.g. low meat consumption, abstinence from alcohol, restricted use of vitamin and mineral supplements, etc.) and medications to reduce iron absorption (e.g. proton pump inhibitors) appear to be reasonable options for patients with haemochromatosis, they have yet to be evaluated in prospective randomised clinical trials [14].

4.2.4 Iron chelation therapy

In patients with acquired iron overload (e.g. anaemia dependent on transfusion), iron-excess management and management of toxicity due to excess iron with chelation have been shown to lower iron burden and increase survival. Patients with serial serum ferritin levels more than 1000 ng/ml and a total infused red blood cell volume of 120 ml/kg of body weight or higher should be treated with chelation treatment, according to recent consensus recommendations. During chelation therapy, serum ferritin levels should be checked every three months to determine that the medication is effectively lowering iron levels. Deferasirox is cost-effective when compared to standard parenteral iron chelation therapy with deferoxamine, according to cost analyses conducted in the United Kingdom and the United States. This is primarily due to the quality-of-life benefits derived from the simpler and more convenient mode of oral administration. The first results from a phase I/II investigation of deferasirox in HFE-haemochromatosis show that a dosage of 5–10 mg/kg/day is sufficient to decrease iron burden, and a randomised trial comparing deferasirox to phlebotomy is now underway [32].

5. Iron metabolism and its effect on caries, microhardness of tooth and discoloration

Although research on iron salts compounds and iron ions support the cariostatic concept, it is difficult to make definitive statements about iron loss owing to a range of chemicals and additives. In the context of a cariogenic diet, however, it appears that specific drops in iron content have a static effect on caries. In light of the current data, it is likely reasonable to state that if a kid consumes carbohydrates that are utilised by cariogenic bacteria, the cariostatic impact might be calculated based on iron drop intake (especially the form of ferrous sulfate. Ferrous Sulphate affects the most as proven in the literature) [33].

In a case–control study by Schroth et al. which aimed to contrast ferritin and haemoglobin levels between preschoolers with S-ECC and caries-free controls, it was concluded that children with S-ECC (severe early childhood caries) appear to be at significantly greater odds of having low ferritin status compared with caries-free children. Children with S-ECC appear to have significantly lower haemoglobin levels and appear to be at significantly greater odds for iron deficiency when compared with caries-free controls.

In the realm of microhardness, the presence of iron in combination with sucrose has resulted in a decrease in the microhardness changes of cow and human enamel. Furthermore, in both in vitro and in vivo conditions, adding iron to acidic liquids reduces demineralization. There is still debate over the mechanism of action of such an ion and its different forms, and this is a fascinating study subject.

Consumption of iron-rich foods (eggs, vegetables, etc.) tends to promote the bacterial growth that produces colouration which is black in the teeth. It has been shown that children with black pigmentations have more calcium and phosphate in their saliva, which can boost the saliva's buffering qualities and lead to a reduction in the occurrence and prevention rate of decay in the presence of pigmentation. However, the relationship between pigmentation, food, oral flora decay and has yet to be found. The combination of iron and sulphide ions produced by bacteria activity is mainly responsible for the iron drop's colour. To justify no indication of colour change in all consumers, the colour change varies with varied iron drop consumption, which might be connected to the total quantity of iron accessible in each drop, the acidity and drops' capacity to etch the surface of the tooth, any bacterial flora, individual's diet and so on [39].

6. Conclusion

The study of microbial iron metabolism is gaining popularity. Initial research on the subject revealed the many ways in which bacteria get iron, began to unravel the crucial function of iron in bacterial metabolism and revealed the means and demand for precise iron absorption management. Iron's role in bacterial pathogenesis has been well documented, and it is currently taken into account in all investigations of prokaryotic pathogens. Basic investigations using *E. coli* and its relatives have given way to studies of less known, and more difficult to grow, organisms, although still incomplete and giving unexpected discoveries, such as the discovery of glucosylated derivatives of enterobactin. Biogenesis research in magnetotactic bacteria has the potential to identify pathways that govern biomineralisation and give insight into organelle development. The potential biotechnology implications of dissimilatory iron reduction research are also intriguing. Because of the extensive and essential role played by environmental interactions between bacteria and iron, geologists, ecologists, environmental and chemical engineers, and physicists, among other professions, have entered the topic. There is a good chance that numerous exciting new discoveries will be made.

Author details

Chinmayee Dahihandekar* and Sweta Kale Pisulkar
Department of Prosthodontics, Sharad Pawar Dental College and Hospital, Maharashtra, India

*Address all correspondence to: chinmayeead@gmail.com

IntechOpen

References

[1] Munoz M, García-Erce JA, Remacha ÁF. Disorders of iron metabolism. Part 1: Molecular basis of iron homoeostasis. Journal of Clinical Pathology. 2011;**64**(4):281-286

[2] Schroth RJ, Levi J, Kliewer E, Friel J, Moffatt ME. Association between iron status, iron deficiency anaemia, and severe early childhood caries: A case–control study. BMC Pediatrics. 2013;**13**(1):1-7

[3] Boyer E, Le Gall-David S, Martin B, Fong SB, Loréal O, Deugnier Y, et al. Increased transferrin saturation is associated with subgingival microbiota dysbiosis and severe periodontitis in genetic haemochromatosis. Scientific Reports. 2018;**8**(1):1-3

[4] Bauminger E, Ofer S, Gedalia I, Horowitz G, Mayer I. Iron uptake by teeth and bones: A Mossbauer effect study. Calcified Tissue International. 1985;**37**(4):386-389

[5] Camaschella C, Poggiali E. Inherited disorders of iron metabolism. Current Opinion in Pediatrics. 2011;**23**(1):14-20

[6] Aisen P, Wessling-Resnick M, Leibold EA. Iron metabolism. Current Opinion in Chemical Biology. 1999;**3**(2):200-206

[7] Camaschella C, Nai A, Silvestri L. Iron metabolism and iron disorders revisited in the hepcidin era. Haematologica. 2020;**105**(2):260

[8] Xi R, Wang R, Wang Y, Xiang Z, Su Z, Cao Z, et al. Comparative analysis of the oral microbiota between iron-deficiency anaemia (IDA) patients and healthy individuals by high-throughput sequencing. BMC Oral Health. 2019; **19**(1):1-3

[9] Lin HP, Wang YP, Chen HM, Kuo YS, Lang MJ, Sun A. Significant association of hematinic deficiencies and high blood homocysteine levels with burning mouth syndrome. Journal of the Formosan Medical Association. 2013;**112**(6):319-325

[10] Wang YP, Chang JY, Wu YC, Cheng SJ, Chen HM, Sun A. Oral manifestations and blood profile in patients with thalassemia trait. Journal of the Formosan Medical Association. 2013;**112**(12):761-765

[11] Wu YC, Wang YP, Chang JY, Cheng SJ, Chen HM, Sun A. Oral manifestations and blood profile in patients with iron deficiency anemia. Journal of the Formosan Medical Association. 2014;**113**(2):83-87

[12] Asgari I, Soltani S, Sadeghi SM. Effects of iron products on decay, tooth microhardness, and dental discoloration: A systematic review. Archives of Pharmacy Practice. 2020;**1**(1):60

[13] Houari S, Picard E, Wurtz T, Vennat E, Roubier N, Wu TD, et al. Disrupted iron storage in dental fluorosis. Journal of Dental Research. 2019;**98**(9): 994-1001

[14] Al Wayli H, Rastogi S, Verma N. Hereditary hemochromatosis of tongue. Oral Surgery, Oral Medicine, Oral Pathology, Oral Radiology, and Endodontology. 2011;**111**(1):e1-e5

[15] Turner J, Parsi M, Badireddy M. Anemia. Treasure Island (FL): StatPearls; 2020

[16] Patton LL, Glick M. The ADA Practical Guide to Patients with Medical conditions. 2nd ed.Hoboken ed. New Jersey: John Wiley & Sons; 2016. pp. 81-88

[17] Derossi SS, Raghavendra S. Anemia. Oral Surgery, Oral Medicine, Oral Pathology, Oral Radiology, and Endodontology. 2003;**95**(2):131-141

[18] Usuki K. Anemia: From basic knowledge to up-to-date treatment. Topic: IV. Hemolytic anemia: Diagnosis and treatment. Nihon Naika Gakkai Zasshi. 2015;**104**(7):1389-1396

[19] Engebretsen KV, Blom-Høgestøl IK, Hewitt S, Risstad H, Moum B, Kristinsson JA, et al. Anemia following Roux-en-Y gastric bypass for morbid obesity; a 5-year follow-up study. Scandinavian Jpurnalof Gastro-enterology. 2018;**53**(8):917-922. DOI: 10.1080/00365521.2018.1489892. 6

[20] Johnson-Wimbley TD, Graham DY. Diagnosis and management of iron deficiency anemia in the 21st century. Therapeutic in Advanced Gastro-enterology. 2011;**4**(3):177-184. DOI: 10.1177/1756283X11398736

[21] Menaa F. Stroke in sickle cell anemia patients: A need for multidisciplinary approaches. Atherosclerosis. 2013;**229**(2):496-503

[22] Mahan LK, Raymond JL. Krause's Food & the Nutrition Care Process. 14th ed. St Louis, Missouri: Elsevier; 2017. pp. 631-643

[23] Wonkam A, Chimusa ER, Mnika K, Pule GD, Ngo Bitoungui VJ, Mulder N, et al. Genetic modifiers of long-term survival in sickle cell anemia. Clinical Translational Medicine. 2020;**10**(4): e152

[24] Helmi N, Bashir M, Shireen A, Ahmed IM. Thalassemia review: Features, dental considerations and management. Electron Physician. 2017;**9**(3):4003-4008. DOI: 10.19082/4003

[25] Karakas S, Tellioglu AM, Bilgin M, Omurlu IK, Caliskan S, Coskun S. Craniofacial characteristics of Thalassemia major patients. Eurasian Journal of Medicine. 2016;**48**(3):204-208. DOI: 10.5152/eurasianjmed.2016.150013

[26] Konda M, Godbole A, Pandey S, Sasapu A. Vitamin B12 deficiency mimicking acute leukemia. Proceedings (Bayl Univ Med Cent). 2019;**32**(4): 589-592

[27] Al-Awami HM, Raja A, Soos MP. Physiology, Gastric Intrinsic Factor. Treasure Island, FL: StatPearls; 2020

[28] Chan CQ, Low LL, Lee KH. Oral vitamin B12 replacement for the treatment of pernicious anemia. Frontiers in Medicine (Lausanne). 2016;**3**:38. DOI: 10.3389/fmed.2016.00038

[29] Linder L, Tamboue C, Clements JN. Drug-Induced vitamin B12 deficiency: A focus on proton pump inhibitors and histamine-2 antagonists. Journal of Pharmaceutical Practise. 2017;**30**(6): 639-642. DOI: 10.1177/08971900 16663092

[30] Damião CP, Rodrigues AO, Pinheiro MF, da Cruz Filho RA, Cardoso GP, Taboada GF, et al. Prevalence of vitamin B12 deficiency in type 2 diabetic patients using metformin: A cross-sectional study. Sao Paulo Medicine Journal. 2016;**134**(6): 473-479. DOI: 10.1590/1516-3180.2015.01382111

[31] Green R, Datta MA. Megaloblastic anemias: Nutritional and other causes. Medical in Clinical North America. 2017;**101**(2):297-317. DOI: 10.1016/j.mcna.2016.09.013

[32] Moore CA, Adil A. Macrocytic Anemia. Treasure Island, FL: StatPearls; 2020

[33] Drexler B, Zurbriggen F, Diesch T, Viollier R, Halter JP, Heim D, et al. Very long-term follow-up of aplastic anemia treated with immunosuppressive therapy or allogeneic hematopoietic cell transplantation. Annals of Hematology. 2020;**99**(11):2529-2538. DOI: 10.1007/s00277-020-04271-4

[34] Tichelli A, de Latour RP, Passweg J, Knol-Bout C, Socié G, Marsh J, et al. SAA Working Party of the EBMT. Long-term outcome of a randomized controlled study in patients with newly diagnosed severe aplastic anemia treated with antithymocyte globulin and cyclosporine, with or without granulocyte colony-stimulating factor: A Severe Aplastic Anemia Working Party Trial from the European Group of Blood and Marrow Transplantation. Haematologica. 2020;**105**(5):1223-1231. DOI: 10.3324/haematol.2019.222562

[35] Vaht K, Göransson M, Carlson K, Isaksson C, Lenhoff S, Sandstedt A, et al. Incidence and outcome of acquired aplastic anemia: Real-world data from patients diagnosed in Sweden from 2000-2011. Haematologica. 2017;**102**(10):1683-1690. DOI: 10.3324/haematol.2017.169862

[36] Cascio MJ, DeLoughery TG. Anemia: Evaluation and diagnostic tests. Medical in Clinical North America. 2017;**101**(2):263-284

[37] Chekroun M, Chérifi H, Fournier B, Gaultier F, Sitbon IY, Ferré FC, et al. Oral manifestations of sickle cell disease. British Dental Journal. 2019;**226**(1):27-31. DOI: 10.1038/sj.bdj.2019.4

[38] Borhade MB, Kondamudi NP. Sickle Cell Crisis. Treasure Island, FL: StatPearls; 2020

[39] McCord C, Johnson L. Oral manifestations of hematologic disease.

Atlas Oral Maxillofacatory Surgery in Clinical North America. 2017;**25**(2):149-162. DOI: 10.1016/j.cxom.2017.04.007

Chapter 6

Ferroptosis: Can Iron Be the Downfall of a Cell?

Asuman Akkaya Fırat

Abstract

Ferroptosis is one of the forms of programmed cell death. Besides being a necessary micronutrient, iron is the key element that initiates ferroptosis in the cell. Intracellular unstable iron accumulation increases the amount of intracellular ROS, especially by the peroxidation of unsaturated membrane phospholipids. Insufficient antioxidant capacity and decreased glutathione levels play an important role in this process. The research reveals that an imbalance between unoxidized polyunsaturated fatty acids (PUFAs) and oxidized PUFAs, particularly oxidized arachidonic acid, accelerates ferroptosis. These oxidative reactions change the permeability of lysosomal and cellular membranes and cell death occurs. Iron chelators, lipophilic antioxidants, and specific inhibitors prevent ferroptosis. In addition to being accepted as a physiological process, it seems to be associated with tissue reperfusion damage, ischemic, neurodegenerative diseases, hematological and nephrological disorders. Ferroptosis is also being explored as a treatment option where it may offer a treatment option for some types of cancer. In this section, the brief history of ferroptosis, its morphological, molecular, and pathophysiological features are mentioned. Ferroptosis seems to be a rich field of research as a treatment option for many diseases in the future.

Keywords: ferroptosis, iron, RCD (regulated cell death), ROS (reactive oxygen species), lipid peroxidation (LP)

1. Introduction

Iron is an essential micronutrient for all living cells. Microorganisms, plant, animal, and human cells need iron to sustain their vital reactions. However, iron overload can cause various metabolic problems and even cell death. Ferroptosis, which has been revealed in the past years, is a form of regulated cell death that develops depending on the increase in the iron load in the cell [1, 2].

Recently, many new cell death modalities have been described. All cell death is considered primarily "regulated cell death" (RCD) and "accidental cell death" (ACD). The accidental cell death occurs in the cell exposed to chemical and physical attacks, independent of genetic coding and molecular pathways. Accidental cell death is a biologically uncontrolled process. Whereas regulated cell death requires signaling cascades and biologically effector molecules. RCD includes apoptosis, necroptosis, autophagy, ferroptosis, pyroptosis, entotic cell death, netotic cell death, parthanatos, lysosome-dependent cell death, alkaliptosis and oxeiptosis [1].

IntechOpen

RDC was first observed in the dying cells of frogs by Karl Vogt in 1842 [1]. Kerr coined the term "apoptosis" for the first time in 1972. Kerr et al. defined apoptosis as a form of programmed cell death (PCD) with morphological changes that differ from necrosis [2]. A new milestone was the identification of CED9 (also known as BCL2 in mammalian cells) and CED4 (also known as apoptotic peptidase activating factor 1 [APAF1] in mammalian cells) from Caenorhabditis elegans development studies in the 1990s [1, 3–5]. RCD is also known as PCD when it occurs in physiological conditions [6]. Apoptosis is considered one of the main reference forms when examining cell death models. Thus, studies on apoptosis accelerated and it was revealed that it can develop in two different ways as intrinsic and extrinsic apoptosis. In recent years, many research results have been revealed about other sub-titles of RCD. The current classification system of cell death has been updated by the Nomenclature Committee on Cell Death (NCCD), which formulates guidelines for the definition and interpretation of all aspects of cell death since 2005 [7].

Ferroptosis is a new type of programmed cell death. In 2003, Dolma et al. discovered the molecule erastin (ST), which has a selective lethal effect on cancer cells that express the RAS family of small GTPases (HRAS, NRAS, and KRAS) protein. The pattern of cell death induced by erestin was different from the previous ones. This new form of cell death did not show any nuclear morphological changes, DNA fragmentation, or caspase activation. In addition, this process was irreversible with caspase inhibitors [8]. Yang and Yagoda found the RSL3 (Ras select and lethal 3) component that inhibits this cellular death pattern and revealed that this cell death process can be stopped by iron chelators [8–10]. Erastin and RSL3 treatment do not induce morphological changes consistent with apoptosis, such as cleavage of ADP-ribose polymerase (PARP). The mechanism of cell death induced by erastin and RSL3 is not attenuated by deletion of the intrinsic apoptotic effectors BCL-2-associated X protein (BAX) and a small molecule inhibitor of BCL-2 antagonist/killer 1 (BAK). These differences distinguish the newly described cell death mechanism from apoptosis, autophagy, and necroptosis. Furthermore, neither mitochondrial ROS production nor Ca^{+2} influx is required for cell death in ferroptosis to occur. Erastin has also been found to cause mitochondrial dysfunction by affecting voltage-dependent anion channels (VDAC) [10]. The term ferroptosis was first used by Dixon et al. For cell death in cancer cells with RAS mutations in 2012. This newly recognized form of cell death can be initiated by iron accumulation and prevented by iron-binding chelators. That's why it's called ferroptosis [8–11].

2. Morphological features of ferroptosis

Ferroptosis is characterized as a cell death model defined as morphologically, reduced mitochondrial volume, decreased or completely absent mitochondrial cristae, increased bilayer membrane density, while the cell membrane is intact, the nucleus remains normal in size, and there is no increase in chromatin density [10, 11]. On electron microscopy, it looks similar to the typical dysmorphic mitochondrial appearance caused by Erastin treatment [12]. Biochemically, intracellular glutathione (GSH) depletion, decreased activity of glutathione peroxidase 4 (GPX4) enzyme, inability to metabolize lipid peroxides, and accumulation of large amounts of ROS (Reactive Oxygen Space) due to iron initiate a lethal process similar to Fenton's reaction and genetically regulated by many genes that have not yet been elucidated [13]. Cancer cells with highly active RAS-RAF-MEK (Receptor Tyrosine Kinases) pathways are susceptible to ferroptosis. The genetic mechanisms that regulate ferroptosis may be related to iron homeostasis and lipid peroxidation [14]. Ferroptosis shows similarities to pathways in other RDC types. Iron-dependent lipid peroxide accumulation is considered to be the basis of the ferroptosis mechanism. It is thought to be a physiological process in mammals rather than a disease or pathological process [15, 16].

3. Accumulation of lipid peroxidase

Mainly phosphatidylethanolamine-OOH (PE-OOH), lipid peroxides are reduced to appropriate lipid alcohols (PE-OOH) by antioxidant reductase mechanisms in the cell under physiological conditions. The effect that will initiate ferroptosis is either by increasing lipid peroxides or by inhibiting the reduction pathway. Glutathione (GSH), the cofactor of glutathione peroxidase (GPX4), is important for the conversion of toxic lipid peroxides to nontoxic lipid alcohols. Glutathione is a tripeptide containing selenocysteine, glutamine, tryptophan. GPX4 catalyzes the following reaction:

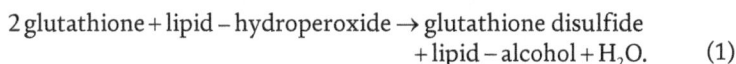

$$2\,glutathione + lipid - hydroperoxide \rightarrow glutathione\ disulfide \\ + lipid - alcohol + H_2O. \qquad (1)$$

This reaction occurs in selenocysteine within the catalytic center of GPX4. During the catalytic cycle of GPX4, active selenol ($-$SeH) is oxidized by peroxides to selenic acid ($-$SeOH) and then reduced by glutathione (GSH) to an intermediate selenodisulfide ($-$Se-SG). GPX4 is eventually reactivated by a second glutathione molecule and glutathione disulfide (GS-SG) is released [17, 18]. GPX4 contains eight neutrophilic amino acids. One of them is selenocysteine and seven cysteines. Selenium, together with cysteine, is essential for the function of GPX4. Inactivation of GPX4 is the most important factor in increasing intracellular lipid ROS and initiating ferroptosis [17–19]. Firstly, in the 1950s, Harry Eagle et al. reported that amino acids, vitamins, and other nutrients are required to protect against oxidative stress in cell culture [20, 21]. Among the molecules reported to be essential was cystine, the oxidized form of cysteine-containing thiol groups [21]. Banni et al., despite glutathione deficiency in human diploid fibroblast cell culture, were able to induce cell growth with α-tocopherol (vitamin E), a lipophilic antioxidant [22].

Compounds that stimulate ferroptosis via GPX4 are divided into four groups (i.e., erastin, RSL3, FIN56, FINO2).

3.1 Erastin

The first group includes erastin (ST). ST inhibits Xc (System Xc-cystine/glutamate antiporter) and decreases intracellular glutathione (GSH) levels. System Xc is an amino acid anti-transporter commonly found in phospholipid bilayer phospholipid membranes. It is an important part of the antioxidant system in cells. It has a heterodimer structure and consists of two subunits, SLC7A11 and SLC3A2, linked to each other by disulfide bonds. System Xc operates on a sodium-free, chlorine-dependent basis. It exchanges cysteine and glutamate in a 1/1 ratio dependent on ATP [21]. Inhibition of the Xc_ system reduces the uptake of cystine, the oxidized form of cysteine [23]. Cysteine is used in intracellular glutathione synthesis [24]. In cells, GSH synthase and glutamate cysteine synthase synthesize GSH with glutamate, glycine, and cysteine, which is reduced from cystine in the cell as substrates [25]. Glutathione reduces the increased load of ROS and the amount of reactive nitrogen decreased cystine causes a decrease in cysteine and depletion of GSH which uses it as a cofactor, to convert lipid peroxides into suitable lipid alcohols, and an increased intracellular oxidant load [25]. Glutathione reduction and decreased glutathione peroxidase activity increase ROS accumulation, oxidative damage, and ultimately ferroptosis [19, 24, 25]. Erastin's blocking of the Xc system and disrupting the intracellular lipid ROS balance damages all intracellular organic substances (e.g. proteins, lipids and nucleic acids), particularly lipid peroxidation initiates ferroptosis. It has been shown that erastin, a prototype compound that inhibits GPX4 via system Xc, also causes ferroptosis by affecting voltage-dependent anion channels. Early chemoproteomic studies showed that voltage-dependent mitochondrial voltage-dependent anion channels 2 and 3 (VDAC2, VDAC3) are direct targets for erastin blockade. VDCA2, purified and reconstituted as artificial liposomes, has been shown to be the target of erastin and modulates transport flow [10, 26]. However, it is accepted that erastin initiates ferroptosis mainly by blocking the X c system cystine/glutamate antitransporter [27]. Also, butionine sulfoximine (BSO), sorafenib, and artesunate induce ferroptosis by depletion of GSH [14, 28, 29]. Reagents or treatments that increase the intracellular amount of cystine/cysteine can reverse erastin-induced ferroptosis, such as β-mercaptoethanol (β-ME), transsulfuration, and processes that increase cysteine synthesis [30–32].

3.2 RSL3

RAS-selective lethal 3 (RSL3), contains an electrophilic moiety and a chloroacetamide moiety and reacts with selenocysteine in the nucleophilic eight amino acid moiety of GPX4, and the enzyme is blocked [33]. Altretamine, which is thought to have a mechanism similar to RSL3, has been defined by the FDA as an anti-cancer drug, but the mechanism of altretamine GPX4 resistance has not been clarified yet [34].

3.3 FIN56

It was named CIL 56, which causes death by ferroptosis in RAS cells while caspas 3 and 7 have no activity. The effect of CIL 56 causing cell death could only be eliminated with low doses of anti-oxidants and iron chelators. At high doses, the lethal effect was irreversible. Later found a CIL56 analog FIN56 (ferroptosis inducing 56) which preserves ferroptosis selectivity in RAS cells. The toxic small-molecule FIN56 is required for mevalonate pathway-mediated ferroptosis. FIN56 can activate its own target protein SQS besides inducing ferroptosis by decreasing the abundance of GPX4 [35]. FIN56 can cause ferroptosis in two ways, either by causing degradation of GPX4 or by reducing the amount of CoQ (i.e., an antioxidant in the cell). The enzymatic activity of acetyl-CoA carboxylase (ACC) is required for FIN56 to degrade GPX4. Therefore, the mechanism of FIN56-induced ferroptosis involves two distinct pathways in association with the mevalonate pathway and fatty acid synthesis. FIN56-mediated mevalonate pathway reduces CoQ. FIN56 binds and activates SQS, the enzyme that converts farnesyl pyrophosphate (FPP) to squalene, which ultimately reduces the level of coenzyme Q10' by reducing the FPP pool available for protein prenylation and metabolite synthesis [35].

3.4 FINO2

FINO2 (endoperoxide-containing 1,2-dioxolane) is a 1,2-dioxolan with both an endoperoxide moiety and a hydroxyl head, capable of inducing ferroptosis. Although its mechanism has not been fully resolved, it indirectly reduces the activity of GPX4. It also provides lipid peroxidation by forming oxygen-centered radicals, similar to the Fenton reaction. Ferroptosis initiated by both FIN56 and FINO2 is partially reversible by β-mercaptoethanol (β-ME) [36, 37].

4. Lipid peroxides and ROS increase

Lipids are important organic molecules because they provide energy for the cell and participate in the structure of cell membranes. Oxygenation of phospholipid (PL) (e.g. PE, phosphatidylcholine, cardiolipin) facilitates ferroptosis. Lipid peroxides are produced in three different ways, each requiringan iron molecule [24, 33, 38]. 1. Lipid ROS produced non-enzymatically by the Fenton reaction with the iron molecule. 2. Lipid peroxides formed by esterification and oxidation of polyunsaturated fatty acids (PUFAs) [24, 33, 39, 40]. 3. Lipid peroxides are formed by catalyzing the iron molecule by lipid auto-oxidation [41]. Fenton reaction is an inorganic reaction that occurs commonly in nature. However, although not fully resolved, (PUFAs) are likely to be the reaction that most contributes to the ferroptosis process [12].

Kagan et al. used RSL3, known as a selective inhibitor, to induce GPX4 inhibition in mouse embryogenic fibroblast cells. RSL3 caused a marked decrease in the chemical activity of GPX4. Among the eight different forms of GPX, GPX4 is the only one that reduces PL-OOH (phospholipid hydroxyls) and PUFA (polyunsaturated fatty acid hydroxyls) in membranes. Kagan et al. screened 350 species of PLs (phospholipids) and identified oxidized AA-containing PE (acyls-arachidonoyl phosphatidyl ethanol) as a ferroptotic cell death signal. AA is a type of PUFAs that can be elongated into adrenoyl (AdA) by elongase [24]. Accumulation of oxidized AA-PE and AdA-PE causes ferroptosis in cells.

It was revealed that the molecule that induces ferroptosis is AA-OOH-PE rather than PL-OOH. The formation of AA-OOH-PE from AA in the cell requires three types of enzymes: 1. lipoxygenases (LOXs), 2. acyl-CoA synthetase long-chain family 4 (ACSL4), and 3. lysophosphatidylcholine acyltransferase 3 (LPCAT3) [11, 17, 24, 39, 40]. In this process, AA is first converted to AA-CoA by being catalyzed by ACSL4, then esterified with LPCAT3 to AA-PE, and finally to AA-OOH-PE a with AA-PE LOXs. Generally accepted views 1. Lipid autoxidation is definitely associated with ferroptosis. 2. Lipid oxidation is associated with ferroptosis rather than lipid peroxidation and is a continuation of lipid peroxidation that cannot be prevented from continuing. 3. Lipid peroxidation initiates lipid autoxidation, while lipid autooxidation causes cell death [38, 41]. In cells undergoing ferroptosis, arachidonic acid (AA) is the most affected by autoxidation. Abundant AA residues were observed in the supernatant of mouse embryo fibroblast (MEFs) with GPX4 depletion. Acyl-CoA synthetase long-chain family4 (ACSL4) and lysophosphatidylcholine acyltransferase3 (LPCAT3) encode enzymes involved in the insertion of AA into membrane phospholipids [42, 43]. ACLSs are composed of proteins expressed on the outer membrane of the endoplasmic reticulum and mitochondria. ACSLs are responsible for the formation of acyl-Cos from fatty acids. There are 5 isoforms: ACLS1, ACSL3, ACSL4, ACSL5, ACSL6 [41]. It has been reported that ACLS4 correlates with ferroptosis. ACLS4 is required for ferroptosis to occur in cells with GPX4 knockout or cells with GPX4 [39]. ACSL4 is not the only enzyme that can activate AA arachidonic acid (arachidonic acid) and ADA, but very high concentrations of AA and AdA are required for ACSL3 to activate, for example. However, AA is normally present in the cell in lower amounts than other fatty acids. Thus, ACSL4 is considered a major regulator for AA and ferroptosis [39]. The mentioned 3 enzymes (ACSL4, LPCAT3, LOX) are active in the initial phase of ferroptosis. As a result of these reactions, the amount of intracellular LOOH increases. Increased intracellular LOOH levels and low valent metals (Fe^{+2}) initiate lipid autooxidation, which is essential for ferroptosis. Lipid autoxidation is the specific and final stage of ferroptosis. Lipid auto-oxidation can only be reversed by radical-trapping antioxidants (RTAs). Lipid autoxidation, rather than LOXs-directed lipid peroxidation, is considered to be the final phase of ferroptosis that causes cell death [38]. It is accepted that intracellular RTA and LOXs levels determine the sensitivity of cells to ferroptosis. According to this assumption, sensitivity to ferroptosis is a physiological process that is affected by many variables such as cell type, physiological conditions and environmental factors [44].

5. Iron and ROS

Iron is an essential element for almost all living organisms. An adult human body contains about 3–5 g of iron. Iron in erythrocytes accounts for 80% of the total iron, and less than 20% is stored in macrophages and hepatocytes. The iron in the macrophage comes from aged red blood cells and is reused, providing 90% of the daily required iron. Approximately 1 g of iron from the diet per day is absorbed through the gastrointestinal tract as 'new iron'. Daily iron loss occurs mostly with the desquamation of epithelial cells in the skin and gastrointestinal tract. In women, menstruation and labor bleeding can cause large amounts of iron losses. The excessive increase of iron in the human body causes hemochromatosis, and less than an adequate amount causes anemia [45, 46]. Inside the cell, iron exists in two forms, Fe^{+2} and Fe^{+3}. The Fe^{+2} form is more functional because of its ability to transfer electrons and have high solubility. The Fe^{+3} form is more stable chemically, so this form is suitable for storage and transportation. Fe^{+2} plays an important role in oxidation-reduction reactions. Fe^{+2} reacts with H_2O_2 to form hydroxyl reagent and Fe^{+3}. Thus, the ROS load inside the cell increases and an oxidative process begins for lipids, proteins, and nucleic acids [47]. It has been accepted that iron and ROS may be increased, especially in tumor cells. The increased oxidative capacity of tumor cells may be effective in their growth. However, it is a contradictory opinion that increased ROS and iron content can increase ferroptosis. Toyokuni et al. reported the hypothesis that intracellular iron and thiol redox groups in tumor cells establish a balance for the cell to avoid ferroptosis. It is recognized that there are many more questions to be answered [25, 48]. Circulating non-heme iron can be transported bound and unbound to transferrin (Tf). Transferrin (Tf) is a glycoprotein with two high-affinity sites specific to ferric iron (Fe^{+3}). When circulating Tf is fully saturated, iron can be transported independently of Tf ferric iron is reduced to ferro (Fe^{+2}) iron by the presence of membrane-bound ferri reductases and taken into the cell by divalent metal transporter 1 (DMT1) [25, 49]. Most of the dietary heme iron is in the form of ferric iron. The absorption of inorganic iron from the lumen into the enterocyte in the duodenal villi is regulated in a very complex and molecularly controlled system. The first step in absorption is Fe^{+3} reduced to Fe^{+2} by ascorbate-dependent duodenal sitokrom b (DCYTB), a membrane-bound reductase. Ferrous iron is taken up into the enterocyte by DMT1 on the lumen-facing surface of the enterocyte. DMT1 is the most important molecule of nonheme iron intake. The synthesis of both DCYTB and DMT 1 is increased in iron deficiency [49–51]. Although intestinal absorption of heme iron (e.g., red meat) is effective for the human organism, it is by a mechanism that is not yet clearly understood. For heme absorption from the duodenum and upper jejunum, coordination of heme carrier protein 1 (HCP1) and heme responsive gene-1 is required [52, 53].

Under physiological conditions, the high-affinity Tf receptor 1 (TfR1) on the cell surface can bind two Fe(III). The Tf-Fe^{+2}-TfR1 complex is transported into cells via endocytosis to form endosomes. Endosomes release iron from the complex under acidic conditions. Free ferric iron is reduced to ferrous iron and then transported into the cytoplasm by DMT1. Endosomes containing the Apo-Tf-TfR1 complex return to the surface of the cells and ferrous iron becomes part of the labile iron pool (LIP) in preparation for the next recycling [49, 50] Iron inside the cell can be stored in ferritin, transferred via ferroportin (FPN), or used in synthesis pathways [44]. Most of the intracellular iron is used for heme and iron-containing proteins, especially mitochondrial iron-sulfur-containing proteins (Fe-S) and iron-dependent enzymes. FPN is the only molecule known to transport iron out of the cell [54]. Iron transferred out of the cell via FPN is in ferrous form. In this transfer, ferrous iron is oxidized by extracellular ferroxidase and converted to ferric iron. The free ferric iron that becomes free is bound to Tf, forming $Tf\text{-}Fe^{+2}$ complexes, and iron is transported to other cells. The iron that is not transported out of the cell and not used in the cell is stored by binding to ferritin. Ferritin is a heterodimer consisting of 24 subunits as ferritin heavy chain 1 (FTH1) and ferritin light chain (FTL). FTH is the domain that binds iron molecules, and FTL can play a role in electron transport. FTH can bind 4500 iron atoms in the ferric form [55]. Iron release from ferritin is also controlled under physiological conditions [56]. In recent studies, nuclear receptor coactivator 4 (NCOA4)-mediated ferrophagy has been shown to induce iron release from ferritin. NCOA4 binds to ferritin and transports it to the lysosome, where ferric iron is decomposed and released [57]. The amount of iron in the cell increases. Therefore, it is postulated that NCOA4-mediated ferritinophagy can induce ferroptosis in the cell [58–61].

6. Regulation of systemic iron

Systemic iron level regulation is carried out through the FPN, which ensures the removal of iron from the cell. FRN is regulated both dependently and independently of hepcidin. In response to adequate systemic iron content, the liver secretes hepcidin into the systemic circulation. Hepcidin binds to FPN in the cell, causing a conformational change in the molecule. The modified FPN molecule is phosphorylated and ubiquitinized, then transported to the lysosome and inactivated by lysosomal enzymes. FPN can also be regulated independently of hepcidin. When there is not enough iron in the cell, FNP undergoes a similar conformational change and is again inactivated in the lysosome. In both cases, the removal of iron from the cell by the FPN is prevented [62].

7. Regulation of intracellular iron

Cell internal iron homeostasis is regulated by iron regulatory protein 1, 2 (IRP1, IRP2) and iron-responsive elements (IREs) molecules. IRPs are proteins that can bind 5',3' (UTR) of the mRNAs of IREs. These proteins are those involved in iron uptake (e.g., DMT1, TfR1), iron sequestration (e.g., subunits of ferritin: FTH1, FTL), and iron export (e.g., FPN). When intracellular iron is deficient, IRPs bind to 5' IREs of ferritin and FPN, and the translocation of these proteins is inhibited [63, 64]. Once iron demand is met, IRPs are degraded and these bonds are removed [63, 64]. Both cellular and systemic iron regulation is related to meeting iron needs. Systemic regulation is provided by the liver and hepcidin. The loss of binding activity of IRP1-IREs for cellular level iron regulation is related to the addition of 4Fe-4S. An E3 ligase complex linked to F-box and leucine-rich repeat protein 5 (FBXL5) drives ubiquitination and proteasomal degradation of IRP-2. FBXL5 requires sufficient iron and oxygen to remain stable. Other genes involved in iron metabolism are IREB2, FBXL5, TfR1, FTH1, and FTL [63–66].

It has been stated that the increase in iron level in the systemic circulation and intracellularly in vivo and in vivo conditions increases the susceptibility to ferroptosis. FPN decreases while Tf increases in ferroptosis-sensitive cells [58, 67]. Lysosomal high

	Ferroptosis
Morphological features	Shrink mitochondria within ceased mitochondrial membrane densities, outer mitochondrial membrane rupture, reduction or vanished of mitochondrial crista, normal nucleus
Biochemical features	Iron accumulation and lipid peroxidation
Death stimulus	Inhibition of cystineimport (e.g.: erastin, SAS, glutamate Glutathione depletion *GPX4* inactivation (e.g. BSO) A.A. depletion in presence of serum and glucose
Regulatory pathway	*Xc/GPX4* pathway, *P53/SLC7 A11* pathway, *ATG5, ATG7-NCOA4* pathway, *MVA* sulfur transfer pathway, *P53-SAT1-ALOX 15* pathway, *HSPB1-TRF1, FSP1-COQ10-NAD(P) H* pathway, *P62-Keap-1NRF2* pathway
Hallmarks	Increased lipid peroxidation iron dependence
Keynes	*GPX4, TFR1, SLC7A11, NRF2, NCOA4, P53, HSPB1, ACSL4, FSP1*

A.A., amino acid; ACSL 4, acyl-Co Asynthetase long-chain family member 4; ALOX-15, arachidonate lipoxygenase 15; AP-1, activator protein-1; ATG5, autophagy-related 5; ATG7, autophagy-related 7; COQ10, coenzyme Q10; DRAM3, damage regulated autophagy modulator 3; FSP1, ferroptosis suppressor protein 1; GPX4, glutathione peroxidase 4; HSPB1, heat shock protein beta-1; Keap1, Keleh-like ECH-associated protein 1; MAPK, mitogen-activated protein kinase; MLKL, mixed lineage kinase domain-like protein; m-TOR, mammalian target of rapamycin; MVA, mevalonate; LC3, microtubule-associated protein 1 light chain 3; NCOA4, nuclear coactivator 4; NRF2, nuclear factor erythroid 2-related factor 2; PKC, protein kinase C; RIP, receptor-interacting serine/threonine kinase; ROS, reactive oxygen species; SAT1, spermidine/spermine N1-acetyltransferase 1; SLC7A11, solute carrier family 7 member 11; system, Xc-cysteine/glutamate transporter; TFEB, transcription factor EB; TFR1, transferrin receptor 1; TNF-R1, tumor necrosis factor R1.

Table 1.
Characteristics of ferroptosis.

concentrations of iron can also prepare cells for ferroptosis [67]. Increased susceptibility to ferroptosis has been observed experimentally in mice with a high amount of iron in their diet and with increased extracellular matrix iron levels [68]. The heat shock protein family B member 1 (HSPB1) inhibits iron uptake via TfR1, reduces the level of iron into the cell, and inhibits ferroptosis by increasing the reduced form of GSH. HSPB1 also inhibits endocytosis and Trf1 reuptake by stabilizing the cortical actin cytoskeleton [69–71]. Both oxygenase 1 (HO-1) and phosphorylase kinase catalytic subunit gamma 2 (PHKG2) mediate ferroptosis when the intracellular iron level is increased [9, 72]. Almost all of the studies on this subject emphasize increased intracellular iron levels and ROS load to initiate ferroptosis. Until now, 3 basic processes that increase ROS in the cell due to iron have been reported: 1. by the Fenton reaction, which is inorganically not enzyme-catalyzed 2. via lipid autooxidation catalyzed by iron-containing enzymes 3. ROS formed by AA oxidation via iron-containing LOXs. However, how iron initiates and maintains ferroptosis and the process leading to cell death has not been fully elucidated (**Table 1**) [44].

8. P62 and NRF2 in ferroptosis

Nuclear factor erythroid 2-associated factor 2 (NRF2) is one of the proteins that create the most important antioxidant response in the cell against oxidative imbalance. Under normal conditions, it is preserved by Kelch-like ECH (erythroid cell-derived protein with CNC homology)-associated protein 1-mediated proteasomal degradation. NRF2 negatively regulates ferroptosis via the p62-keap1-NRF2 pathway. NRF2 and p62 competitively bind to Keap1 [73, 74]. Nrf2 plays a vital role in intracellular antioxidant balancing and activation of GPX4, in the re-synthesis of NADPH, 6PGD (phospho-gluconate dehydrogenase, malic enzyme, and glucose 6-phosphate dehydrogenase) and glutathione synthesis, cysteine supply via system Xc (glutathione peroxidase 4, glutathione reductase) play a key role for many genes. Ferroptosis inducers facilitate the interaction between p62 and Keap1. This interaction inhibits Keap1. Inhibition of Keap1 prevents binding between Keap1 and NRF2. The interaction of Keap1 and NRF2 triggers the degradation of NRF2 [75, 76]. It leads to NRF2-mediated ferroptosis by downregulating genes involved in iron and ROS metabolisms. The most important of these are such as quinone oxidoreductase 1 (NQO1 and HO1) [75, 76].

NRF2 inhibits ferroptosis by increasing the expression of target genes involved in iron and ROS metabolism, such as NQO1 (NADPH Quinone Dehydrogenase 1) and HO1 (Heme oxygenase 1). NRF2-Keap1 pathway supports system Xc so NRF2 inhibits ferroptosis. It also negatively regulates ferroptosis by lowering intracellular reactive iron by gene regulation of Nrf2 ferritin (FTL/FTH) light chain and heavy chains, ferroportin (SLC40A1) subunit, and SLC7A11 component of Xc system. Nrf2 is also activated by oxidized lipids, which are also involved in the initiation of ligand-mediated transcription factor PPAR γ (peroxisome proliferator-activated receptor-gamma). Furthermore, a high NRF2 expression is associated with a worse overall survival rate in patients with glioma, and activation of the NRF2-Keap1 pathway supports system Xc so NRF2 inhibits ferroptosis [77]. The existence of studies reporting that NRF2 induces ferroptosis shows that there are still unanswered questions on this subject.

9. Tumor suppressor protein P53 and ferroptosis

P53 is a tumor suppressor that has been extensively studied. It has a tumor-suppressive effect by stopping metabolic cycles, mediating aging and apoptosis. It is involved in the cellular response to DNA damage, hypoxia, starvation, and oncogene activation. Activation of p53 ensures cell cycle slowdown at the low level of cellular stress, repair DNA damage, prevent ROS accumulation, and cell survival. However, severe cellular stress and damage induce a response of P53 to produce apoptosis and cell death [78]. On the one hand, p53 suppresses ferroptosis either through direct inhibition of DPP4 (dipeptidyl peptidase 4) activity [79] or through induction of CDKN1A/p21 (cyclin-dependent kinase inhibitor 1A) expression. On the other hand, 53 can increase ferroptosis by inhibiting the expression of SLC7A11 (solute transporter family 7 member 11) or by increasing the expression of SAT1 (spermidine/spermine N1-acetyltransferase 1) and GLS2 (glutaminase 2) [78–80].

It is accepted that the direction and intensity of the response of p53 are proportional to the level of stress to which the cell is exposed [80]. It is depleted by a mutation in many types of cancer and its anti-tumor effect is limited [81, 82]. Unlike nuclear p53, which acts as an autophagy-promoting transcription factor [82, 83], cytosolic p53 can block autophagy in response to nutrient starvation or mTOR inhibition [80, 83, 84]. These context-dependent roles of p53 in survival and death are regulated in a fine-tuned manner by its ubiquitination, phosphorylation, acetylation, and other modifications. Unlike nuclear p53, which functions as a transcription factor promoting autophagy, cytotic p53 (BCL-2 family (BAX [BCL2 associated X, apoptosis regulator] and BBC3/PUMA [BCL2 binding component 3]) can suppress autophagy in response to cellular starvation and mTOR inhibition [81–84]. p53 cellular response is regulated by ubiquitination, phosphorylation, acetylation, and other modifications [81].

Classic ferroptosis model and cell culture studies have revealed that P53 is associated with ferroptosis [79, 82, 84]. The researchers have found that 53 promotes ferroptosis due to transrepression of SLC7A11 expression in fibroblasts and some cancer cells (human breast cancer MCF7) and human osteosarcoma (U2OS) [79, 85, 86]. P53 plays a role in ferroptosis cascades, which can cause cell survival or death. It can function prodeath or prosurvival at the transcriptional or post-translational level. Depending on the type or severity of stress that the cell is exposed to, it may contribute to apoptosis or autophagy [87, 88].

Inhibition of SLC7A11 expression, increased expression of SAT1 (spermidine/ Spermine N1-acetyltransferase 1), increased expression of GLS (Glutaminase) are required for p53-regulated ferroptosis [78]. SAT1 is a regulator of polyamine metabolism. Oxidative stress, inflammatory stimuli, and heat shock have been found to stimulate SAT1 activity. SAT1 is a transcriptional target of p53. An increase in SAT 1 does not change SLC7A11 and GPX4 activity but increases ALOX15 (arachidonate 15-lipoxygenase) activity [84]. Thus, the required antioxidant response remains insufficient despite increased lipid peroxidation.

Acetylation of K98 is crucial for p53-mediated ferroptosis. In particular, p53 3KR, an acetylation-defective mutant in which 3 lysine residues (at positions 117, 161, and 162) have been replaced by arginine residues, is highly effective in repressing SLC711A [85–87]. In contrast, p53 4KR98 (an acetylation-defective mutant in which an additional lysine is replaced at position 98) cannot reduce SLC711A expression [80].

Perhaps p53 3KR gains a ferroptosis-inducing capacity while p53 4KR loses it. In human cancers, wild-type p53 is degraded by high levels of the oncogenic E3 ubiquitin-protein ligase MDM2. Thus, inhibition of MDM2-dependent proteasomal degradation of p53 offers an attractive therapeutic strategy for cancer therapy [88]. Since it will not be inactivated in MDM2−/− cells, the p53 level increases. p53 has been shown to contribute to the cell death cascade, which can be termed ferroptosis, which can be reversible by ferrostatin 1 in MDM2−/− mouse embryos. However, another study showed that ferrostatin-1 alone could not prevent cell death caused by MDM2 deficiency [89–91].

The anti-ferroptosis activation of ferrostatin-1 and liproxstatin-1 (another widely-used ferroptosis inhibitor) are mediated through their reactivity as radical-trapping antioxidants rather than their potency as inhibitors of lipoxygenases [90, 91]. The acetylation levels of p53 are localized by six different histone acetyltransferase: 1. REBBP/CBP (CREB binding protein), 2. EP300/p300 (E1A binding protein P300), 3. KAT2B/PCAF (lysine acetyltransferase 2B), 4. KAT5/Tip60 (lysine acetyltransferase), 5. KAT8/MOF (lysine acetyltransferase 8), and 6. KAT6A/MOZ (lysine acetyltransferase 6A). The ability of these acetyltransferases to regulate ferroptosis remains unclear [88, 92, 93].

GSL2 (glutaminase 2) is a mitochondrial enzyme, the first step of glutamine catabolism, and an important regulator of ferroptosis [94]. It is known as a transcriptional target of p53. It is responsible for oxygen consumption and ATP production in cancer cells. It is also known to offer antioxidant support through the production of GSH [95]. While all this is expected for negative regulation of ferroptosis, it has been shown that glutaminase degradation inhibits ferroptosis in fibroblast cells [96]. More research is needed for the relationship between glutaminase, p53, and ferroptosis.

DPP4 (dipeptidyl peptidase-4) is the most important regulator of survival in the ferroptosis-related function of P53. Cells with p53 knockout or pharmacologically inhibited become more sensitive to type I inducer of ferroptosis (erastin and SAS). However, there is no difference in response to typeII ferroptosis inducer (RSL3 and FIN56).

However, DPP4 inhibitors (linagliptin, vildagliptin, and alogliptin) together with other protease inhibitors (doxycycline, ritonavir, atazanavir, VX-222, semagacestat) completely block erastin-induced cell death in p53-deficient cells [79].

Another mediator of p53, CDKN1A/p21 (cyclin-dependent kinase inhibitor 1A), inhibits apoptosis. In cystine deficiency in cancer cells, p53-mediated CDKN1A expression delays ferroptosis. Again, inhibition of MDM2 by nutlin-3 increases expression of p53, which blocks the ferroptosis induced by loss of Xc function [82]. More comprehensive and detailed studies on p53 and ferroptosis are needed.

10. Beclin-1 and ferroptosis

Beclin-1 (Vps30/Atg6 in yeast) is a well-known regulator of autophagy primarily involved in the formation of the PtdIns3K complex, which is involved in activating autophagy. Beclin-1 is a critical regulator of ferroptosis that is independent of the formation of the PtdIns3K complex. The beclin-1 expression only affects ferroptosis induced by the system Xc-inhibitor. Knockdown of Beclin-1 by RNA interference (RNAi) blocks ferroptosis, whereas knockdown of Beclin-1 by gene transfection promotes ferroptosis in cancer cells in response to system Xc-inhibitors (for example, erastin, sulfasalazine, and sorafenib). In contrast, it does not affect erastin-, sorafenib-, or sulfasalazine-induced ferroptosis. Beclin-1 mandatory for ferroptosis induced by system Xc – inhibitor [97, 98].

It needs ATG5 (related to autophagy 5) and NCOA4 (nuclear receptor coactivator 4). ATG5 is part of an E3-like ligase that is critical for the lipidation of members of GABARAP (GABA type-A receptor-associated protein families) and MAP1LC3 (microtubule-associated protein 1 light chain 3) members. However, NCOA4 is a transporter receptor that mediates FT/ferritin degradation via selective ferritinophagy. Inhibits elastin-induced conversion of MAP1LC3B-I to MAP1LC3B-II by inhibition of Atg5. Furthermore, suppression of NCOA4 blocks the degradation of FT/ferritin, resulting in suppression of ferroptosis. In contrast, knockdown of Beclin-1 does not affect the synthesis of lapidated MAP1LC3B and MAP1LC3B-positive points in ferroptosis. As a positive control in starvation-induced cells, knockdown of Beclin-1 stops the conversion of MAP1LC3B-I to MAP1LC3B-II. Significantly, the formation of a BECN1-PtdIns3K complex was observed in cancer cells only in response to starvation, but not to ferroptotic stimulus. These findings point to the regulatory roles of Beclin-1 in ferroptosis compared to induced autophagy [96–98].

11. AMPK ferroptosis

AMP-activated protein kinase (AMPK), a critical indicator of the cell's energy deficit, is activated through AMP binding, kinase phosphorylation, and other mechanisms. AMP-activated protein kinase (AMPK), a critical indicator of the cell's energy deficit, is activated through AMP binding, kinase phosphorylation, and other mechanisms. AMPK maintains the viability of the cell under energy stress. If this energy balance cannot be achieved, it leads the cell to apoptosis. AMPK exhibits various regulatory roles in lipid metabolism by mediating the phosphorylation of acetyl-CoA carboxylase as well as polyunsaturated fatty acid biosynthesis. AMPK has also been implicated in ferroptosis. The inhibitory effect of AMPK activation on ferroptosis does not include modulation of cystine uptake, iron metabolism autophagy, or mTORC1 signaling. Energy stress-mediated AMPK activation inhibits ferroptosis via mitochondria. The Loss of function of liver kinase B1 (LKB1) sensitizes mouse embryonic fibroblasts (MEFs) and human non-small cell lung carcinoma cell lines to ferroptosis. This LKB1-AMPK-ACC1 (ACC1—Acetyl-CoA carboxylase 1)-FAS (cell surface death receptor) axis has a vital role in regulating ferroptotic cell death [99].

A recent study also reported a supportive role of AMPK in the regulation of Beclin-1-mediated ferroptosis. Specifically, AMPK mediates the phosphorylation of Beclin-1 at Ser90/93/96. This is a prerequisite for the formation of the Beclin1-SLC7A11 complex in ferroptosis and subsequent lipid peroxidation. Inhibition of AMPK by siRNA or compound C reduces erastin-induced Beclin-1 phosphorylation at S93/96, thus inhibiting the formation of Beclin-1-SLC7A11 complex formation and subsequent ferroptosis. Thus, it is clear that Beclin-1 contributes to the core molecular machinery and signaling pathways involved in ferroptosis [100, 101]. The mechanisms of AMPK-mediated regulatory ferroptosis need further investigation.

12. Ataxia-telangiectasia-mutated kinase in ferroptosis

Ataxia-telangiectasia mutated kinase (ATM) is a crucial kinase for DNA damage responses. P53 is one of its sub-targets, which plays a decisive role in the regulation of ferroptosis, which activates its role in ferroptosis [102]. Genetic or pharmacological inhibition of ATM reduces intracellular labile iron by increasing FPN, FTH1, and FTL. Ataxia-telangiectasia mutated kinase (ATM) is a crucial kinase for DNA damage responses. P53 is one of its sub-targets, which plays a decisive role in the regulation of ferroptosis, which activates its role in ferroptosis [103]. Genetic or pharmacological inhibition of ATM reduces intracellular labile iron by increasing FPNand FTH and FTL. It relies on the transcriptional activity and nuclear translocation of metal regulatory transcription factor 1 (MTF1) upon TM inhibition. Under conditions of ATM inhibition, nuclear translocation of MTF1 is increased, resulting in changes in ferritin (FTH1) and ferroportin (FPN) expression, and the amount of intracellular unstable iron is reduced to prevent ferroptosis [104].

13. Iron and ferroptosis

The iron homeostasis in both manners is regulated by iron. In systemic iron regulation, the level of iron is sensed by the liver and the liver secretes the hormone hepcidin according to iron abundance. At the cellular iron level, the loss of IRP1-IREs binding activity depends on the insertion of 4Fe–4S cluster. As for the IRP2, a newly discovered FBXL5-dependent E3 ligase complex catalyzes the ubiquitination and proteasomal degradation of IRP2, while keeping the stability of FBXL5 requires iron and oxygen IREB2, the main regulator of iron metabolism upon inhibition, reduces sensitivity to ferroptosis. Since iron metabolism is also affected by autophagy, it also regulates ferroptosis in many ways [58]. Ferritinophagy is the autophagy of selective ferritin, in which ferritin is recognized by the specific transport receptor NCOA4, which directs it to autophagosomes for lysosomal degradation. This lysosomal degradation of ferritin releases iron and thus increases ferroptosis susceptibility [45].

Apart from ferritin, HSPB1 and CISD1 are other proteins that affect ferroptosis susceptibility. In addition, heme oxygenase 1 (HO-1) and phosphorylase kinase catalytic subunit gamma 2 (PHKG2) mediate ferroptosis by regulating the abundance of iron [9, 72]. ROS accumulation initiated by labile iron in the cell occurs in three known ways, respectively. 1. non-enzymatic Fenton reaction; 2. ROS accumulated by lipid autooxidation catalyzed by iron-containing enzymes; 3. ROS accumulated by oxidation of arachidonic acid (AA) by lipid peroxidases (LOX). Although it is known that iron is an essential element in ferroptosis, it is not fully understood how iron regulates ferroptosis.

14. Biomarkers of ferroptosis

Prostaglandin-endoperoxide synthase 2 (PTGS2) increases with the formation of lipid peroxides and decreases in nicotinamide adenine dinucleotide phosphate (NADPH) [9, 32]. Moreover, the increase in PTGS2 cannot be suppressed by PTGS inhibitors. Malondialdehyde (MDA), the end product of lipid peroxidation, also increases. GPX4 protects cells against ferroptosis by catalyzing GSH and toxic PE-AA-OOH to oxidize GSH (GSSG) and non-toxic PE-AA-OH. GSSG is then converted to GSH by GSH reductase (GR) in the presence of NADPH. Therefore, NADPH, a coenzyme of GR, plays a vital role in maintaining the abundance of intracellular GSH. Furthermore, the basal NADPH abundance of a given cell is negatively correlated with ferroptosis susceptibility. NADPH necroptosis can establish a link between ferroptosis and GSH (**Figure 1**) [9, 32].

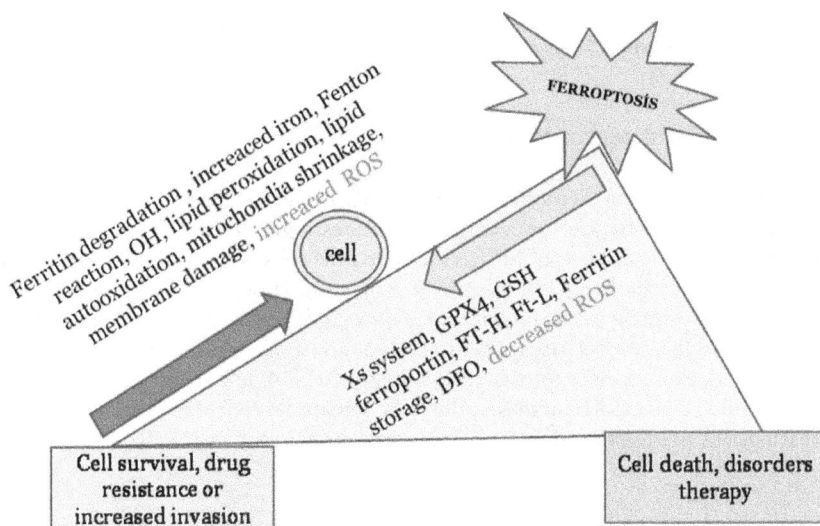

Figure 1.
ROS (reactive oxygen space) and ferroptosis.

15. Ferroptosis and other forms of cell death

Unlike ferroptosis, apoptosis, necrosis, and autophagy decreased mitochondrial volume, increased mitochondrial membrane density, reduction in mitochondrial cristae, and rupture of the outer membrane are observed. Also, ferroptosis cannot be stopped by inhibitors of apoptosis, necrosis, and autophagy [8, 9, 27]. Compared to ferroptosis, it may show common features with other regulated cell deaths. Although ferroptosis does show mitochondrial differences, it cannot be entirely attributed to it. The amount of mitochondrial ROS does not change in ferroptosis exposed to erastin. Moreover, ferroptosis also occurs in cells lacking a mitochondrial electron transport chain [8, 14, 105].

16. Ferroptosis and oxytosis

Oxidative glutamate toxicity leads to glutathione depletion by inhibiting cystine uptake via exogenous glutamate, the Xc system (cystine/glutamate antiporter).

Activation of lipoxygenase opens up cGMP-gated channels that allow reactive oxygen species production and extracellular calcium influx. In oxytosis paradigms in neuronal cells, mitochondrial disorders are mediated through mitochondrial transactivation of pro-apoptotic protein (BID). Upon translocation to mitochondria, BID mediates loss of mitochondrial integrity and function and deleterious translocation of mitochondrial AIF to the nucleus. Induced by stress and independently of mitochondrial death. In neuronal cells, ROS-induced transactivation of BID into mitochondria links both oxytosis and ferroptosis pathways and leads to irreversible morphological and functional damage. MMP (Matrix metallo proteinases) loss, reduction of ATP levels, and mitochondrial ROS generation are associated with the apoptosis-inducing factor (AIF). The BID inhibitor BI-6c9 and the ferroptosis inhibitors ferrostatin-1 and liproxstatin-1 can block these deadly pathways upstream of mitochondrial disorders. The analogy of ferroptosis and oxytosis seems to be the most promising for treatment options, especially in diseases related to iron accumulation such as neurodegenerative diseases, stroke, and reperfusion injury [106–108].

17. Ferroptosis and necroptosis

They used ACSL4 as a ferroptosis susceptibility marker and a Mixed Lineage Kinase domain-like (MLKL) marker for necroptosis. Interestingly, ACSL4 deficiency led to an increase in MLKL, and loss of MLKL increased the cells' sensitivity to ferroptosis. When one cell death pathway is inhibited, it evolves into the other pathway. It has been demonstrated that ferroptosis and necroptosis are different forms of cell death. They used ACSL4 as a ferroptosis susceptibility marker and a mixed lineage kinase domain-like (MLKL) marker for necroptosis. Interestingly, ACSL4 deficiency led to an increase in MLKL, and loss of MLKL increased the cells' sensitivity to ferroptosis. When one cell death pathway is inhibited, it evolves into the other pathway [15, 16, 109, 110].

18. Ferroptosis and autophagy

Ferritinophagy is the autophagic process of ferritin mediated by NCOA4. NCOA4 binds to FTH1 in autophagosomes during low intracellular iron and then autophagosomes are sent to lysosomes for ferritin degradation. Similarly, autophagy is the death of the cell by breaking down organelles and proteins in the cell with phagosomes and lysosomes [57]. Cell death can also be blocked by the inhibition of ferritinophagy in aged cells [60]. Ferritinophagy and unstable iron increase in fibroblast culture and cancer cells accelerated cell death [59]. Increased ferritinophagy in liver fibrosis and erastin-induced ferroptosis follow the same process [61]. Ferritinophagia has been reported at the onset of ferroptosis [100]. However, it differs from ferritinophagy in that specific autophagy inhibitors fail to rescue ferroptosis [11].

19. Conclusion

Ferroptosis differs from other forms of regulated cell death. It requires an unstable form of intracellular iron. It is associated with the increased reactive oxidative load. The susceptibility to ferroptosis differs from cell to cell. It presents new research areas for the treatment of cancer, circulatory diseases, and degenerative neurological disorders.

Abbreviations

AA	acyls-arachidonoyl
ACC1	acetyl-CoA carboxylase 1
ACD	accidental cell death
ACSL4	acyl-CoA synthetase long-chain family member 4
AIF	apoptosis-inducing factor
ALOX-15	arachidonate lipoxygenase 15
AMPK	AMP-activated protein kinase
APAF1	apoptotic peptidase 1
ATG5	autophagy-related 5
ATG7	autophagy-related 7
ATM	ataxia-telangiectasia mutated kinase
BAX	BCL-2 associated protein
BBC3/PUMA	BCL2 binding component 3
BCL-2	B-cell lymphoma 2 gene
β-ME	β-mercaptoethanol
BSO	butionine sulfoximine
CDKN1A/p21	cyclin-dependent kinase inhibitor 1A
CED9	cell death abnormality gene 9
COQ10	Coenzyme Q10
DCYTB	ascorbate-dependent duodenal sitokrom *b*.
DPP4	dipeptidyl peptidase-4
DRAM3	demage regulated autophagy modulator 3
FBXL5	F-box and leucine-rich repeat protein 5
FIN56	ferroptosis inducing 56
FINO2	ferroptosis inducer endoperoxide containing 1,2-dioxolane
FPN	ferroportine
FSP1	ferroptosis suppressor protein 1
FTH	ferritin heavy chain
FTL	ferritin light chain
FPP	farnesyl pyrophosphate.
GABARAP	GABA type-A receptor-associated protein families
GPX4	glutathion peroxydase 4
GSH	glutathione
HO1	heme oxygenase
HSPB1	H shock protein beta-1
IREs	responsive elements
IRPs	iron regulatory proteins
KAT2B/PCAF	lysine acetyltransferase 2B/P300/CBP-associated factor
LC3	microtubule-associated protein 1 light chain3
LIP	labile iron pool
LKB1	liver kinase B1
LOXs	lipoxygenases
MAPK	mitogen-aktivated protein kinase
MAP1LC3	microtubule-associated protein 1 light chain 3
MEF	mouse embrionic fibroblast
MMP	matrix metalloproteinases
MLKL	mixed lineage kinase domain-like protein
MTF1	metal regulatory transcription factor 1

mTOR	mammalian target of rapamycin
MVA	mevalonate
NRF2	nuclear factor erythroid 2-associated factor 2
NADPH	nicotinamide adenine dinucleotide phosphate
NCCD	nomenclature committee on cell death
NCOA4	nuclear coactivator4
NQO1	NADPH quinone dehydrogenase1
NRF2	nuclear factor erythroid 2-related factor 2
PCD	programmed cell death
PE-OOH	phosphatidyl ethanol amine-hydroxyl
PHKG2	phosphorylase kinase catalytic subunit gamma 2
PKC	protein kinase C
PL	phospholipid
PTGS2	prostaglandin-endoperoxide synthase 2
PUFAs	polyunsaturated fatty acids
RAS-RAF-MEK	receptor tyrosine kinases
RCD	regulated cell death
REBBP/CBP	CREB binding protein
RIP	receptor-interacting serine/threonine kinase
ROS	reactive oxygen species
RSL3	RAS-selective lethal 3
RTAs	radical-trapping antioxidants
SAT1	spermidine/spermine N1-acetyltransferase 1
SLC7A11	solute carrier family 7 member 11
SQS	squalene synthase
VDAC	voltage-dependent anion channels
Tf	transferrin
TfR1	Tf receptor 1
TFEB	transcription factor EB
TfR1	transferrin receptor 1
TNF-R1	tumor necrosis factor R1
U-2 OS	cell line human osteosarcoma
Xc system	cystine/glutamate antiporter system

Author details

Asuman Akkaya Fırat
Biochemistry Department, Fatih Sultan Mehmet (F.S.M) Training and Research
Hospital, Health Sciences University (S.B.U), İstanbul, Turkey

*Address all correspondence to: asumanfirat44@gmail.com

IntechOpen

References

[1] Tang D, Kang R, Berghe TV, et al. The molecular machinery of regulated cell death. Cell Research. 2019;**29**:347-364. DOI: 10.1038/s41422-019-0164-5

[2] Kerr JF, Wyllie AH, Currie AR. Apoptosis: A basic biological phenomenon with wide-ranging implications in tissue kinetics. British Journal of Cancer. 1972;**26**(4):239-257. DOI: 10.1038/bjc.1972.33

[3] Hengartner M, Ellis R, Horvitz R. *Caenorhabditis elegans* gene *ced-9* protects cells from programmed cell death. Nature. 1992;**356**:494-499. DOI: 10.1038/356494a

[4] Hengartner MO, Horvitz HR. *C. elegans* cell survival gene ced-9 encodes a functional homolog of the mammalian proto-oncogene bcl-2. Cell. 1994;**76**(4):665-676. DOI: 10.1016/0092-8674(94)90506-1

[5] Yuan J, Horvitz HR. The *Caenorhabditis elegans* cell death gene ced-4 encodes a novel protein and is expressed during the period of extensive programmed cell death. Development. 1992;**116**(2):309-320

[6] Galluzzi L, Bravo-San Pedro J, Vitale I, et al. Essential *versus* accessory aspects of cell death: Recommendations of the NCCD 2015. Cell Death and Differentiation. 2015;**22**:58-73. DOI: 10.1038/cdd.2014.137

[7] Galluzzi L, Vitale I, Aaronson S, et al. Molecular mechanisms of cell death: Recommendations of the Nomenclature Committee on Cell Death 2018. Cell Death and Differentiation. 2018;**25**:486-541. DOI: 10.1038/s41418-017-0012-4

[8] Dolma S, Lessnick SL, Hahn WC, Stockwell BR. Identification of genotype-selective antitumor agents using synthetic lethal chemical screening in engineered human tumor cells. Cancer Cell. 2003;**3**(3):285-296. DOI: 10.1016/s1535-6108(03)00050-3

[9] Yang WS, Stockwell BR. Synthetic lethal screening identifies compounds activating iron-dependent, nonapoptotic cell death in oncogenic-RAS-harboring cancer cells. Chemistry & Biology. 2008;**15**:234-245

[10] Yagoda N, von Rechenberg M, Zaganjor E, et al. RAS-RAF-MEK-dependent oxidative cell death involving voltage-dependent anion channels. Nature. 2007;**447**:864-868. DOI: 10.1038/nature05859

[11] Dixon SJ et al. Ferroptosis: An iron-dependent form of nonapoptotic cell death. Cell. 2012;**149**:1060-1072

[12] Cao JY, Dixon SJ. Mechanisms of ferroptosis. Cellular and Molecular Life Sciences. 2016;**73**(11-12):2195-2209. DOI: 10.1007/s00018-016-2194-1

[13] Li J, Cao F, Yin H, et al. Ferroptosis: Past, present and future. Cell Death & Disease. 2020;**11**:88. DOI: 10.1038/s41419-020-2298-2

[14] Friedmann Angeli JP, Schneider M, Proneth B, et al. Inactivation of the ferroptosis regulator Gpx4 triggers acute renal failure in mice. Nature Cell Biology. 2014;**16**(12):1180-1191. DOI: 10.1038/ncb3064

[15] Linkermann A, Skouta R, Himmerkus N, Mulay SR, Dewitz C, De Zen F, et al. Synchronized renal tubular cell death involves ferroptosis. Proceedings of the National Academy of Sciences of the United States of America.

2014;**111**:16836-16841. DOI: 10.1073/pnas.1415518111

[16] Zille M, Karuppagounder SS, Chen Y, Gough PJ, Bertin J, Finger J, et al. Neuronal death after hemorrhagic stroke *in vitro* and *in vivo* shares features of ferroptosis and necroptosis. Stroke. 2017;**48**:1033-1043. DOI: 10.1161/strokeaha.116.015609

[17] Yang WS, SriRamaratnam R, Welsch ME, Shimada K, Skouta R, Viswanathan VS, et al. Regulation of ferroptotic cancer cell death by GPX4. Cell. 2014;**156**:317-331. DOI: 10.1016/j.cell.2013.12.010

[18] Kinowaki Y, Kurata M, Ishibashi S, Ikeda M, Tatsuzawa A, Yamamoto M, et al. Glutathione peroxidase 4 overexpression inhibits ROS-induced cell death in diffuse large B-cell lymphoma. Laboratory Investigation. 2018;**98**:609-619. DOI: 10.1038/s41374-017-0008-1

[19] Jiang L, Hickman JH, Wang SJ, Gu W. Dynamic roles of p53-mediated metabolic activities in ROS-induced stress responses. Cell Cycle. 2015;**14**(18):2881-2885. DOI: 10.1080/15384101.2015.1068479

[20] Eagle H. Nutrition needs of mammalian cells in tissue culture. Science. 1955;**122**:501-514. DOI: 10.1126/science.122.3168.501

[21] Eagle H, Piez KA, Oyama VI. The biosynthesis of cystine in human cell cultures. The Journal of Biological Chemistry. 1961;**236**:1425-1428

[22] Bannai S, Tsukeda H, Okumura H. Effect of antioxidants on cultured human diploid fibroblasts exposed to cystine-free medium. Biochemical and Biophysical Research Communications. 1977;**74**:1582-1588. DOI: 10.1016/0006-291X(77)90623-4

[23] Sato H, Tamba M, Ishii T, Bannai S. Cloning and expression of a plasma membrane cystine/glutamate exchange transporter composed of two distinct proteins. The Journal of Biological Chemistry. 1999;**274**(17):11455-11458. DOI: 10.1074/jbc.274.17.11455

[24] Kagan VE, Mao G, Qu F, Angeli JP, Doll S, Croix CS, et al. Oxidized arachidonic and adrenic PEs navigate cells to ferroptosis. Nature Chemical Biology. 2017;**13**(1):81-90. DOI: 10.1038/nchembio.2238

[25] Toyokuni S, Ito F, Yamashita K, Okazaki Y, Akatsuka S. Iron and thiol redox signaling in cancer: An exquisite balance to escape ferroptosis. Free Radical Biology & Medicine. 2017;**108**:610-626. DOI: 10.1016/j.freeradbiomed.2017.04.024

[26] Bauer AJ, Gieschler S, Lemberg KM, et al. Functional model of metabolite gating by human voltage-dependent anion channel 2. Biochemistry. 2011;**50**:3408-3410. DOI: 10.1021/bi2003247

[27] Dixon SJ, Patel DN, Welsch M, et al. Pharmacological inhibition of cystine-glutamate exchange induces endoplasmic reticulum stress and ferroptosis. eLife. 2014;**3**:e02523. DOI: 10.7554/eLife.02523

[28] Lachaier E, Louandre C, Godin C, Saidak Z, Baert M, Diouf M, et al. Sorafenib induces ferroptosis in human cancer cell lines originating from different solid tumors. Anticancer Research. 2014;**34**:6417-6422

[29] Eling N, Reuter L, Hazin J, Hamacher-Brady A, Brady NR. Identification of artesunate as a specific activator of ferroptosis in pancreatic cancer cells. Oncoscience. 2015;**2**(5):517-532. DOI: 10.18632/oncoscience.160

[30] Ishii T, Bannai S, Sugita Y. Mechanism of growth stimulation of

L1210 cells by 2-mercaptoethanol in vitro. Role of the mixed disulfide of 2-mercaptoethanol and cysteine. The Journal of Biological Chemistry. 1981;**256**:12387-12392

[31] Hayano M, Yang WS, Corn CK, Pagano NC, Stockwell BR. Loss of cysteinyl-tRNA synthetase (CARS) induces the transsulfuration pathway and inhibits ferroptosis induced by cystine deprivation. Cell Death and Differentiation. 2016;**23**(2):270-278. DOI: 10.1038/cdd.2015.93

[32] Shimada K, Stockwell BR. tRNA synthase suppression activates de novo cysteine synthesis to compensate for cystine and glutathione deprivation during ferroptosis. Molecular & Cellular Oncology. 2015;**3**(2):e1091059. DOI: 10.1080/23723556.2015.1091059

[33] Yang WS, Kim KJ, Gaschler MM, Patel M, Shchepinov MS, Stockwell BR. Peroxidation of polyunsaturated fatty acids by lipoxygenases drives ferroptosis. Proceedings of the National Academy of Sciences of the United States of America. 2016;**113**(34): E4966-E4975. DOI: 10.1073/pnas.1603244113

[34] Woo JH, Shimoni Y, Yang WS, Subramaniam P, Iyer A, Nicoletti P, et al. Elucidating compound mechanism of action by network perturbation analysis. Cell. 2015;**162**:441-451. DOI: 10.1016/j.cell.2015.05.056

[35] Shimada K, Skouta R, Kaplan A, Yang WS, Hayano M, Dixon SJ, et al. Global survey of cell death mechanisms reveals metabolic regulation of ferroptosis. Nature Chemical Biology. 2016b;**12**:497-503. DOI: 10.1038/nchembio.2079

[36] Abrams RP, Carroll WL, Woerpel KA. Five-membered ring

peroxide selectively initiates ferroptosis in cancer cells. ACS Chemical Biology. 2016;**11**:1305-1312. DOI: 10.1021/acschembio.5b00900

[37] Gaschler MM, Andia AA, Liu H, Csuka JM, Hurlocker B, Vaiana CA, et al. FINO2 initiates ferroptosis through GPX4 inactivation and iron oxidation. Nature Chemical Biology. 2018a;**14**:507-515. DOI: 10.1038/s41589-018-0031-6

[38] Shah R, Shchepinov MS, Pratt DA. Resolving the role of lipoxygenases in the initiation and execution of ferroptosis. ACS Central Science. 2018;**4**:387-396. DOI: 10.1021/acscentsci.7b00589

[39] Doll S, Proneth B, Tyurina YY, Panzilius E, Kobayashi S, Ingold I, et al. ACSL4 dictates ferroptosis sensitivity by shaping cellular lipid composition. Nature Chemical Biology. 2017;**13**:91-98. DOI: 10.1038/nchembio.2239

[40] Shintoku R, Takigawa Y, Yamada K, Kubota C, Yoshimoto Y, Takeuchi T, et al. Lipoxygenase-mediated generation of lipid peroxides enhances ferroptosis induced by erastin and RSL3. Cancer Science. 2017;**108**:2187-2194. DOI: 10.1111/cas.13380

[41] Soupene E, Fyrst H, Kuypers FA. Mammalian acyl-CoA:lysophosphatidylcholine acyltransferase enzymes. Proceedings. National Academy of Sciences. United States of America. 2008;**105**:88-93. DOI: 10.1073/pnas.0709737104

[42] Soupene E, Kuypers FA. Mammalian long-chain acyl-CoA synthetases. Experimental Biology and Medicine (Maywood, N.J.). 2008;**233**: 507-521. DOI: 10.3181/0710-MR-287

[43] Shindou H, Shimizu T. Acyl-CoA:lysophospholipid acyltransferases. The Journal of Biological Chemistry.

2009;**284**:1-5. DOI: 10.1074/jbc. R800046200

[44] Lei P, Bai T, Sun Y. Mechanisms of ferroptosis and relations with regulated cell death: A review. Frontiers in Physiology. 2019;**10**:139. DOI: 10.3389/fphys.2019.00139

[45] Muckenthaler MU, Galy B, Hentze MW. Systemic iron homeostasis and the iron-responsive element/iron-regulatory protein (IRE/IRP) regulatory network. Annual Review of Nutrition. 2008;**28**:197-213. DOI: 10.1146/annurev.nutr. 28.061807.155521

[46] Lawen A, Lane DJ. Mammalian iron homeostasis in health and disease: Uptake, storage, transport, and molecular mechanisms of action. Antioxidants & Redox Signaling. 2013;**18**:2473-2507. DOI: 10.1089/ars.2011.4271

[47] Pignatello JJ, Oliveros E, MacKay A. Advanced oxidation processes for organic contaminant destruction based on the fenton reaction and related chemistry. Critical Reviews in Environmental Science and Technology. 2006;**36**:1-84. DOI: 10.1080/10643380500326564

[48] Kasai H. Analysis of a form of oxidative DNA damage, 8-hydroxy-2'-deoxyguanosine, as a marker of cellular oxidative stress during carcinogenesis. Mutation Research. 1997;**387**(3):147-163. DOI: 10.1016/s1383-5742(97)00035-5

[49] Richardson DR, Ponka P. The molecular mechanisms of the metabolism and transport of iron in normal and neoplastic cells. Biochimica et Biophysica Acta. 1997;**1331**:1-40. DOI: 10.1016/s0304-4157(96)00014-7

[50] Hentze MW, Muckenthaler MU, Galy B, Camaschella C. Two to tango: Regulation of mammalian iron metabolism. Cell. 2010;**142**:24-38. DOI: 10.1016/j.cell.2010.06.028

[51] Lane DJ, Bae DH, Merlot AM, Sahni S, Richardson DR. Duodenal cytochrome b (DCYTB) in iron metabolism: An update on function and regulation. Nutrients. 2015;**7**(4):2274-2296. DOI: 10.3390/nu7042274

[52] Shayeghi M, Latunde-Dada GO, Oakhill JS, Laftah AH, Takeuchi K, Halliday N, et al. Identification of an intestinal heme transporter. Cell. 2005;**122**:789-801. DOI: 10.1016/j.cell.2005.06.025

[53] Rajagopal A, Rao AU, Amigo J, Tian M, Upadhyay SK, Hall C, et al. Haem homeostasis is regulated by the conserved and concerted functions of HRG-1 proteins. Nature. 2008;**453**:1127-1131. DOI: 10.1038/nature06934

[54] Donovan A, Brownlie A, Zhou Y, Shepard J, Pratt SJ, Moynihan J, et al. Positional cloning of zebrafish ferroportin1 identifies a conserved vertebrate iron exporter. Nature. 2000;**403**:776-781. DOI: 10.1038/3500156

[55] Arosio P, Levi S. Cytosolic and mitochondrial ferritins in the regulation of cellular iron homeostasis and oxidative damage. Biochimica et Biophysica Acta. 2010;**1800**:783-792. DOI: 10.1016/j.bbagen.2010.02.005

[56] Kurz T, Terman A, Gustafsson B, Brunk UT. Lysosomes in iron metabolism, ageing and apoptosis. Histochemistry and Cell Biology. 2008;**129**:389-406. DOI: 10.1007/s00418-008-0394-y

[57] Mancias JD, Wang X, Gygi SP, Harper JW, Kimmelman AC. Quantitative proteomics identifies NCOA4 as the cargo receptor mediating ferritinophagy. Nature. 2014;**509**:105-109. DOI: 10.1038/nature13148

[58] Gao M, Monian P, Quadri N, Ramasamy R, Jiang X. Glutaminolysis

and transferrin regulate ferroptosis. Molecular Cell. 2015;**59**:298-308. DOI: 10.1016/j.molcel.2015.06.011

[59] Hou W, Xie Y, Song X, Sun X, Lotze MT, Zeh HJ III, et al. Autophagy promotes ferroptosis by degradation of ferritin. Autophagy. 2016;**12**:1425-1428. DOI: 10.1080/15548627.2016.1187366

[60] Masaldan S, Clatworthy SAS, Gamell C, Meggyesy PM, Rigopoulos AT, Haupt S, et al. Iron accumulation in senescent cells is coupled with impaired ferritinophagy and inhibition of ferroptosis. Redox Biology. 2018;**14**:100-115. DOI: 10.1016/j.redox.2017.08.015

[61] Zhang Z, Yao Z, Wang L, Ding H, Shao J, Chen A, et al. Activation of ferritinophagy is required for the RNA-binding protein ELAVL1/HuR to regulate ferroptosis in hepatic stellate cells. Autophagy. 2018;**14**:2083-2103. DOI: 10.1080/15548627.2018.1503146

[62] Sangkhae V, Nemeth E. Regulation of the iron homeostatic hormone hepcidin. Advances in Nutrition. 2017;**8**:126-136. DOI: 10.3945/an.116.013961

[63] Anderson CP, Shen M, Eisenstein RS, Leibold EA. Mammalian iron metabolism and its control by iron regulatory proteins. Biochimica et Biophysica Acta. 2012;**1823**:1468-1483. DOI: 10.1016/j.bbamcr.2012.05.010

[64] Thompson JW, Bruick RK. Protein degradation and iron homeostasis. Biochimica et Biophysica Acta. 2012;**1823**:1484-1490. DOI: 10.1016/j.bbamcr.2012.02.003

[65] Salahudeen AA, Thompson JW, Ruiz JC, Ma HW, Kinch LN, Li Q, et al. An E3 ligase possessing an iron-responsive hemerythrin domain is a regulator of iron homeostasis. Science. 2009;**326**:722-726. DOI: 10.1126/science.1176326

[66] Vashisht AA, Zumbrennen KB, Huang X, Powers DN, Durazo A, Sun D, et al. Control of iron homeostasis by an iron-regulated ubiquitin ligase. Science. 2009;**326**:718-721. DOI: 10.1126/science.1176333

[67] Ma S, Henson ES, Chen Y, Gibson SB. Ferroptosis is induced following siramesine and lapatinib treatment of breast cancer cells. Cell Death & Disease. 2016;**7**:e2307. DOI: 10.1038/cddis.2016.208

[68] Wang H, An P, Xie E, Wu Q, Fang X, Gao H, et al. Fare hemokromatoz modellerinde ferroptozun karakterizasyonu. Hepatoloji. 2017;**66**:449-465. DOI: 10.1002/hep.29117

[69] Arrigo AP, Virot S, Chaufour S, Firdaus W, Kretz-Remy C, Diaz-Latoud C. Hsp27 consolidates intracellular redox homeostasis by upholding glutathione in its reduced form and by decreasing iron intracellular levels. Antioxidants & Redox Signaling. 2005;**7**:414-422. DOI: 10.1089/ars.2005.7.414

[70] Chen H, Zheng C, Zhang Y, Chang YZ, Qian ZM, Shen X. Heat shock protein 27 downregulates the transferrin receptor 1-mediated iron uptake. The International Journal of Biochemistry & Cell Biology. 2006;**38**:1402-1416. DOI: 10.1016/j.biocel.2006.02.006

[71] Sun X, Ou Z, Xie M, Kang R, Fan Y, Niu X, et al. HSPB1 as a novel regulator of ferroptotic cancer cell death. Oncogene. 2015;**34**:5617-5625. DOI: 10.1038/onc.2015.32

[72] Kwon MY, Park E, Lee SJ, Chung SW. Heme oksijenaz-1, erastin kaynaklı ferroptotik hücre ölümünü hızlandırır. Oncotarget. 2015;**6**:24393-24403. DOI: 10.18632/oncotarget.5162

[73] Sun X, Ou Z, Chen R, Niu X, Chen D, Kang R, et al. Activation of the p62-Keap1-NRF2 pathway protects against ferroptosis in hepatocellular carcinoma cells. Hepatology. 2016;**63**:173-184. DOI: 10.1002/hep.28251

[74] Komatsu M, Kurokawa H, Waguri S, Taguchi K, Kobayashi A, Ichimura Y, et al. The selective autophagy substrate p62 activates the stress responsive transcription factor Nrf2 through inactivation of Keap1. Nature Cell Biology. 2010;**12**:213-223. DOI: 10.1038/ncb2021

[75] Padmanabhan B, Tong KI, Ohta T, Nakamura Y, Scharlock M, Ohtsuji M, et al. Structural basis for defects of Keap1 activity provoked by its point mutations in lung cancer. Molecular Cell. 2006;**21**:689-700. DOI: 10.1016/j. molcel.2006.01.013

[76] Tong KI, Padmanabhan B, Kobayashi A, Shang C, Hirotsu Y, Yokoyama S, et al. Different electrostatic potentials define ETGE and DLG motifs as hinge and latch in oxidative stress response. Molecular and Cellular Biology. 2007;**27**:7511-7521. DOI: 10.1128/ MCB.00753-07

[77] Fan Z, Wirth AK, Chen D, Wruck CJ, Rauh M, Buchfelder M, et al. Nrf2-Keap1 pathway promotes cell proliferation and diminishes ferroptosis. Oncogene. 2017;**6**:e371. DOI: 10.1038/oncsis.2017.65

[78] Kastenhuber ER, Lowe SW. Putting p53 in context. Cell. 2017;**170**(6):1062-1078. DOI: 10.1016/j. cell.2017.08.028

[79] Xie Y, Zhu S, Song X, Sun X, Fan Y, Liu J, et al. The tumor suppressor p53 limits ferroptosis by blocking DPP4 activity. Cell Reports. 2017;**20**(7):1692-1704

[80] Kruiswijk F, Labuschagne CF, Vousden KH. p53 in survival, death and metabolic health: A lifeguard with a licence to kill. Nature Reviews. Molecular Cell Biology. 2015;**16**(7):393-405. DOI: 10.1038/nrm4007

[81] Chipuk JE, Bouchier-Hayes L, Kuwana T, Newmeyer DD, Green DR. PUMA couples the nuclear and cytoplasmic proapoptotic function of p53. Science. 2005;**309**(5741):1732-1735. DOI: 10.1126/science.1114297

[82] Tarangelo A, Magtanong L, Bieging-Rolett KT, Li Y, Ye J, Attardi LD, et al. p53 suppresses metabolic stress-induced ferroptosis in cancer cells. Cell Reports. 2018;**22**(3):569-575. DOI: 10.1016/j.celrep.2017.12.077

[83] Kang R, Kroemer G, Tang D. The tumor suppressor protein p53 and the ferroptosis network. Free Radical Biology & Medicine. 2019;**133**:162-168. DOI: 10.1016/j.freeradbiomed. 2018.05.07

[84] Drakos E, Atsaves V, Li J, et al. Stabilization and activation of p53 downregulates mTOR signaling through AMPK in mantle cell lymphoma. Leukemia. 2009;**23**:784-790. DOI: 10.1038/leu.2008.348

[85] Jiang L, Kon N, Li T, Wang SJ, Su T, Hibshoosh H, et al. Ferroptosis as a p53-mediated activity during tumour suppression. Nature. 2015;**520**(7545):57-62. DOI: 10.1038/nature14344

[86] Ou Y, Wang SJ, Li D, Chu B, Gu W. Activation of SAT1 engages polyamine metabolism with p53-mediated ferroptotic responses. Proceedings of the National Academy of Sciences of the United States of America. 2016;**113**(44):E6806-E6812. DOI: 10.1073/pnas.1607152113

[87] Gao W, Shen Z, Shang L, et al. Upregulation of human autophagy-initiation kinase ULK1 by tumor suppressor p53 contributes to DNA-damage-induced cell death. Cell Death and Differentiation. 2011;**18**:1598-1607. DOI: 10.1038/cdd.2011.33

[88] Reed SM, Quelle DE. p53 acetylation: Regulation and consequences. Cancers (Basel). 2014;**7**(1):30-69. DOI: 10.3390/cancers7010030

[89] Moll UM, Petrenko O. The MDM2-p53 interaction. Molecular Cancer Research. 2003;**1**(14):1001-1008

[90] Zilka O, Shah R, Li B, Friedmann Angeli JP, Griesser M, Conrad M, et al. On the mechanism of cytoprotection by ferrostatin-1 and liproxstatin-1 and the role of lipid peroxidation in ferroptotic cell death. ACS Central Science. 2017;**3**(3):232-243. DOI: 10.1021/acscentsci.7b00028

[91] Thomasova D, Bruns HA, Kretschmer V, Ebrahim M, Romoli S, Liapis H, et al. Murine double minute-2 prevents p53-overactivation-related cell death (podoptosis) of podocytes. Journal of the American Society of Nephrology: JASN. 2015;**26**(7):1513-1523. DOI: 10.1681/asn.2014040345

[92] Wang SJ, Li D, Ou Y, Jiang L, Chen Y, Zhao Y, et al. Acetylation is crucial for p53-mediated ferroptosis and tumor suppression. Cell Reports. 2016;**17**(2):366-373. DOI: 10.1016/j.celrep.2016.09.022

[93] Dai C, Gu W. p53 post-translational modification: Deregulated in tumorigenesis. Trends in Molecular Medicine. 2010;**16**(11):528-536. DOI: 10.1016/j.molmed.2010.09.002

[94] Altman BJ, Stine ZE, Dang CV. From Krebs to clinic: Glutamine metabolism to cancer therapy. Nature Reviews Cancer. 2016;**16**(10):619-634

[95] Hu W, Zhang C, Wu R, Sun Y, Levine A, Feng Z. Glutaminase 2, a novel p53 target gene regulating energy metabolism and antioxidant function. Proceedings of the National Academy of Sciences. 2010;**107**(16):7455-7460

[96] Kim EM, Jung C-H, Kim J, Hwang S-G, Park JK, Um H-D. The p53/p21 complex regulates cancer cell invasion and apoptosis by targeting Bcl-2 family proteins. Cancer Research. 2017;**77**(11):3092-3100

[97] Swaminathan G, Zhu W, Plowey ED. BECN1/Beclin 1 tür hücre yüzeyi APP/amiloid β lizozomal bozulması için ön-madde proteini. Otofaji. 2016;**12**(12):2404-2419

[98] Tang D, Kang R. Regulation and function of autophagy during ferroptosis. In: Ferroptosis in Health and Disease. Springer International Publishing; 2019. pp. 43-59. DOI: 10-1007/978-3-030-26780-3_3

[99] Li C, Dong X, Du W, et al. LKB1-AMPK axis negatively regulates ferroptosis by inhibiting fatty acid synthesis. Signal Transduction and Targeted Therapy. 2020;**5**:187. DOI: 10.1038/s41392-020-00297-2

[100] Gao M, Yi J, Zhu J, Minikes AM, Monian P, Thompson CB, et al. Role of mitochondria in ferroptosis. Molecular Cell. 2019;**73**(2):354-363.e3. DOI: 10.1016/j.molcel.2018.10.042

[101] Song X, Zhu S, Chen P, Hou W, Wen Q, Liu J, et al. AMPK-mediated BECN1 phosphorylation promotes ferroptosis by directly blocking system X_c^- activity. Current Biology.

2018;**28**(15):2388-2399.e5. DOI: 10.1016/j.cub.2018.05.094

[102] Anand SK, Sharma A, Singh N, Kakkar P. Entrenching role of cell cycle checkpoints and autophagy for maintenance of genomic integrity. DNA Repair. 2020;**86**:article 102748

[103] Zhang W, Gai C, Ding D, Wang F, Li W. Targeted p53 on small-molecules-induced ferroptosis in cancers. Frontiers in Oncology. 2018;**8**:507

[104] Chen PH, Wu J, Ding C-KC. Kinome screen of ferroptosis reveals a novel role of ATM in regulating iron metabolism. Cell Death & Differentiation. 2020;**27**(3):1008-1022

[105] Gaschler MM, Hu F, Feng H, Linkermann A, Min W, Stockwell BR. Determination of the Subcellular Localization and Mechanism of Action of Ferrostatins in Suppressing Ferroptosis. ACS Chemical Biology. 2018 Apr 20;**13**(4):1013-1020. DOI: 10.1021/acschembio.8b00199

[106] Landshamer S, Hoehn M, Barth N, Duvezin-Caubet S, Schwake G, Tobaben S, et al. Bid-induced release of AIF from mitochondria causes immediate neuronal cell death. Cell Death and Differentiation. 2008;**15**:1553-1563. DOI: 10.1038/cdd.2008.78

[107] Neitemeier S, Jelinek A, Laino V, Hoffmann L, Eisenbach I, Eying R, et al. BID links ferroptosis to mitochondrial cell death pathways. Redox Biology. 2017;**12**:558-570. DOI: 10.1016/j.redox.2017.03.007

[108] Tobaben S, Grohm J, Seiler A, Conrad M, Plesnila N, Culmsee C. Bid-mediated mitochondrial damage is a key mechanism in glutamate-induced oxidative stress and AIF-dependent cell death in immortalized HT-22 hippocampal neurons. Cell Death and Differentiation. 2011;**18**:282-292

[109] Tonnus W, Linkermann A. "Death is my Heir"–ferroptosis connects cancer pharmacogenomics and ischemia-reperfusion injury. Cell Chemical Biology. 2016;**23**:202-203. DOI: 10.1016/j.chembiol.2016.02.005

[110] Muller T, Dewitz C, Schmitz J, Schroder AS, Brasen JH, Stockwell BR, et al. Necroptosis and ferroptosis are alternative cell death pathways that operate in acute kidney failure. Cellular and Molecular Life Sciences. 2017;**74**:3631-3645. DOI: 10.1007/s00018-017-2547-4

Section 3

Overview on Iron Deficiency

Chapter 7

Potential Marker for Diagnosis and Screening of Iron Deficiency Anemia in Children

Yulia Nadar Indrasari, Siti Nurul Hapsari
and Muhamad Robiul Fuadi

Abstract

Iron plays a role in multiple physiological functions, naming oxygen transport, gene regulation, DNA synthesis, DNA repair, and brain function. Iron deficiency anemia (IDA) may happen following iron deficiency, but iron deficiency alone may cause negative impacts on the health risk of pediatric patients. The degree of iron deficiency is described by total body iron (measured by ferritin), transport iron (measured by transferrin saturation), serum iron, and other hematologic and biochemical markers. Iron deficiency anemia is a result of insufficient iron supply causing the inability to maintain normal levels of hemoglobin. The most common causes of microcytic anemia in children are iron deficiency and thalassemia minor. There are various hematologic and biochemical parameters used for screening and diagnosis of iron deficiency anemia in children, but there is no single "best" test to diagnose iron deficiency with or without anemia. The "gold standard" for identifying iron deficiency is a direct test-bone marrow biopsy with Prussian blue staining. This article aims to explain iron metabolism in children and discuss the role of hematologic and biochemical parameters for screening and diagnosis of iron deficiency anemia in children.

Keywords: iron deficiency anemia, health risk, children, diagnosis, screening

1. Introduction

Iron deficiency is the most common nutritional deficiency across the world and an important public health problem, particularly in developing countries [1]. Anemia, defined as a low hemoglobin concentration, is a public health problem that affects low-, middle-, and high-income countries, having significant adverse health consequences, as well as adverse impacts not only to the health of citizens, but also to the socio-economic development [2]. The high prevalence of iron deficiency anemia in developing countries most often is attributed to nutritional deficiencies worsened by chronic blood loss due to parasitic infections and malaria. In the industrialized nations, the most common cause of iron deficiency with or without anemia is insufficient dietary iron [3].

Approximately 50% of cases of anemia are considered to be an iron deficiency, but the proportion probably varies among population groups and in different areas, according to the local conditions [2]. Unfortified complementary foods particularly have a low iron content, iron deficiency, also iron deficiency anemia (IDA) are consequently major public health problems in infants and young children, especially in poor populations [4].

Anemia resulting from iron deficiency adversely affects cognitive and motor development, causes fatigue and low productivity [2]. Iron plays a role in various essential physiological functions, such as oxygen transport, gene regulation, DNA synthesis, DNA repair, and brain function [5]. Iron serves important functions in biochemical processes including the development of the central nervous system, and it is essential to neural myelination, neurotransmitter function, neuronal energy metabolism and neurite differentiation [6].

Many studies have shown an association between iron deficiency anemia and poor neurodevelopment in infants that lasts beyond the period of deficiency [6]. This article aims to explain iron metabolism in children, also discuss the role of hematologic and biochemical parameters for screening and diagnosis of iron deficiency anemia (IDA) in children.

2. Iron requirements, absorption, and metabolism in infants and children

Hemoglobin levels at birth are normally quite high and primarily consist of fetal hemoglobin (HbF or $\alpha_2\gamma_2$), which comprises 80–90% of the total hemoglobin synthesized, gradually decreasing to <1% by 10 months of age in normal infants. The switch from hemoglobin F to adult hemoglobin (HbA or $\alpha_2\beta_2$) begins around 12 weeks of gestation, although the production of hemoglobin A occurs in the bone marrow where it remains throughout the life [7].

Iron requirements in late infancy are higher than during any other period in life due to rapid growth. A unique feature of human iron metabolism is the absence of an excretory pathway and regulation of iron absorption is very important for homeostasis [4]. At birth, most of the body iron is found in the blood hemoglobin, but a term, healthy, normal birth weight infant also has some iron stores, appropriate to about 25% of the total body iron [8].

Knowledge of iron metabolism in infants and children has been enhanced due to recent discoveries of protein and peptides regulating iron absorption. Iron is absorbed in the small intestine by divalent metal transporter 1 (DMT1) and is stored by ferritin inside the mucosal cells or taken by ferroportin to the systemic circulation, while being oxidized by hephaestin to be integrated into transferrin. Hepcidin, a small peptide that is synthesized by the liver, can sense iron stores and regulates iron transport by ferroportin inhibition [7].

2.1 Iron requirements

The majority of iron required by the body is obtained from the reuse of iron released from erythrocyte catabolism. However, sufficient amounts of iron must be supplied by the diet to replace the iron that is lost from the body (through exfoliation of the skin and gastrointestinal cells; and blood loss) and the iron that is needed for growth [9]. At birth, most of the body iron is found in blood hemoglobin, but a term, healthy, normal birth weight infant also has some iron stores, corresponding to about 25% of the total body iron [8].

Healthy, full-term, normal birth weight infants are born with sufficient stores of iron to cover their needs during the first 4–6 months of life [9]. The healthy infant at term is born with iron stores which can be partially mobilized and utilized for growth during early infancy. In addition to these stores, the high levels of hemoglobin at birth will decrease and the iron will be recycled and also used for growth and blood-volume expansion [7].

The huge demand for iron in the late fetal and early postnatal period is for hemoglobin (Hb) synthesis [10]. Some theories are estimating the iron requirements of infants. Total body iron varies with birthweight and has been estimated to be approximately 268 mg for an infant with a birthweight of 3.5 kg and approximately 183 mg for an infant with a birthweight of 2.5 kg [7].

Premature infants are at high risk of iron deficiency (ID) due to inadequate iron storage caused by the factors of preterm birth, early onset of postnatal erythropoiesis, and rapid growth after birth. There is a lack of a gold standard to describe iron status clinically for healthy preterm infants [10].

2.2 Iron absorption in children and mechanism regulating iron absorption

Iron bioavailability is commonly assumed to be 50% from breast milk and 10% from mixed foods. The stable isotope method can be applied to assess iron absorption in children [4]. Iron homeostasis is mainly controlled through tightly regulated changes in iron absorption in adults. Three "regulators" of iron hemostasis mechanisms have been identified which are referred to as the "erythropoietic regulator", the "stores regulator", and the "dietary regulator" [7].

Iron deficiency and overload are protected by the regulation of these compartments which are integrated to control iron absorption. The store's regulator has a predominant role in maintaining iron homeostasis in response to endogenous iron stores. The dietary regulator may functionally respond to acute changes in iron intake, primarily to prevent iron overload [7].

Iron is absorbed from the diet in primarily the duodenum and jejunum. Iron cannot pass through cellular membrane unassisted. The primary importer of iron across the apical membrane of the intestinal epithelial cell is divalent metal transporter 1 (DMT1, also known as Nramp2, and DCT1). To date, only 1 transmembrane transporter protein, solute carrier family 11, member 2 (Slc11a2, also known as DMT1, is known to have physiological importance in bringing iron into cells. DMT1 is essential for iron absorption, based on a murine study explained that lack the gene encoding DMT1 develop severe IDA. Slc11a2 acts as a proton-dependent iron importer of Fe^{2+}. It can also transport a variety of other divalent metal cations, including Mn^{2+}, Co^{2+}, Cu^{2+} and Zn^{2+} [11].

Iron homeostasis is regulated at the level of intestinal absorption. Several proteins must synchronize the transfer of iron across the enterocyte and into the systemic circulation. Iron acquired from the diet, is generally in the ferric (Fe^{3+}) state and must be reduced to the ferrous form (Fe^{2+}) before uptake into the enterocyte, presumably by an apical membrane-associated ferric reductase, possibly duodenal cytochrome b (Dcytb), aiding the uptake of ferrous iron across the apical membrane into the enterocyte via divalent metal transporter 1 (DMT1) [7]. In adults, the only known transport mechanism for iron from the intestinal lumen into the enterocyte, is DMT1. In the fetal and infant human small intestine, a lactoferrin receptor has been found, and may be important for iron uptake lactoferrin binds most iron in breast milk. The functional importance of the lactoferrin receptor in human infants, is still not yet determined [4].

Hepcidin has an important role in the regulation of iron absorption. The hepatic synthesis of this peptide is induced by high serum iron concentrations and circulating hepcidin leads to decreased expression of ferroportin on the basolateral membrane of enterocytes, thereby blocking the dietary iron transport into the blood. On the contrary, hepcidin is downregulated in iron deficiency leading to an increase in intestinal iron absorption. It is not yet known whether hepcidin is involved in the significant developmental changes in iron metabolism that occur during the first year of life [4].

Iron homeostasis is primarily regulated at the level of intestinal absorption in adults; thus, the ontogeny of this homeostatic system has developmental consequences. The study from Lŏnnerdal and Kelleher explained a hypothesis that the increase in iron absorption that occurs during infancy reflects the maturation of the small intestine iron absorption mechanism to facilitate iron transfer into the systemic circulation [7].

2.3 Iron metabolism

Despite the magnitude of the difference in bioavailability of iron from breast milk and infant formula varies among studies, most investigators agree that iron is absorbed better from breast milk. A major part of iron in breast milk is associated with lactoferrin. Human lactoferrin is absorbed across the apical membrane of the intestinal cell via a specific lactoferrin receptor and internalized with its bound iron. Thus, lactoferrin facilitates a unique mechanism for the absorption of iron from breast milk. The molecular reasons for the lack of homeostasis of iron metabolism in young infants are not yet known. Iron absorption is refractory to hepcidin during the neonatal period, despite intact hepcidin signaling during this period. The mechanism for iron absorption and its regulation is different during early life than in adults, so further research is needed in this area [6].

Transferring transports absorbed iron to the liver, where it is taken up into hepatocytes by transferrin receptors and stored sequestered in ferritin until needed. Iron is released from ferritin and mobilized into the hepatic circulation for further distribution to the tissue, during times of high demand. The regulation of this process is just beginning to be explained, and our concept has been aided by the identification of several genes expressed in the liver that when mutated cause hereditary hemochromatosis, resulting in iron overload. These genes contain those for hepcidin, hemochromatosis protein (HFE), transferrin receptor 2 (TfR2), and hemojuvelin [7].

3. Iron deficiency Anemia (IDA) in children

3.1 Etiology of iron deficiency anemia

Anemia may be caused by decreased RBC production, increased RBC destruction, or blood loss [12]. In developing countries, iron deficiency (ID) and iron deficiency anemia (IDA) typically result from insufficient dietary intake, loss of blood due to intestinal worm colonization, or both. In high-income countries, certain eating habits (e.g., vegetarian diet) and pathologic conditions (e.g., chronic blood loss or malabsorption) are the most common causes [13].

Inadequate intake together with rapid growth, low birth weight and gastrointestinal loss due to excessive consumption of cow's milk are the most common causes of IDA in children. Iron crossing through the placenta is the only source of iron during the intrauterine period. In the final period of pregnancy, the total amount of iron in the fetus is 75 mg/kg. If there is no significant cause of blood loss, physiological anemia begins in the postnatal period and iron stores are sufficient to provide erythropoiesis in the first

6 months of life. Stores are exhausted earlier in babies with perinatal blood loss and in low birth weight infants, since they are smaller. Improvement of the iron status and reduction of the risk of iron deficiency can be done by delaying umbilical cord clamping [1].

The iron-fortified formula helps ensure adequate iron supplies for infants. However, toddlers often have diets that contain large amounts of cow milk and minimal amounts of iron-rich foods. The risk of iron deficiency may be increased by the early introduction of whole cow milk (before 1 year of age) and consumption of greater whole cow milk after the first year of life. Cow milk is not only low in iron, it also interferes with iron absorption. Cow milk may cause unknown gastrointestinal bleeding in some infants [14].

Adolescent females may become anemic due to menstrual losses. Some children develop anemia due to Meckel diverticulum, chronic epistaxis, or inflammatory bowel disease, which all cause blood loss. Iron is absorbed from the gastrointestinal tract and transported into the blood bound to transferrin. Excess iron is stored primarily in the liver, bone marrow, and spleen as ferritin [14].

3.2 Iron deficiency: clinical classification and clinical findings

Three main body iron compartments describe iron status inadequacy: iron stores, transport iron, and functional iron. Depletion of each component leads to a different iron deficiency stage. Short-term variations in physiologic iron needs are met by the release of iron stores, the majority of which are available as intracellular ferritin, predominantly in hepatocytes and specialized macrophages [15].

Iron deficiency (ID) can be divided into 4 major categories: 1) iron depletion (a state in which the low level of iron affects nonhematologic pathways (e.g., brain, muscle); where microcytic anemia that is classically seen in iron deficiency anemia (IDA) is not found, 2) iron-restricted erythropoiesis (a condition with some impairment of hematologic function without evidence of anemia or microcytosis), 3) IDA (a clinical picture with reduced hemoglobin levels, in which neurodevelopmental and musculoskeletal functions have been inhibited), 4) Functional iron deficiency (a state in which iron stores are adequate but unavailable for biological use). This typical laboratory findings of each category can be seen in **Table 1** [5].

Laboratory finding	Iron depletion	Iron-restricted erythropoiesis	Iron deficiency anemia	Functional iron deficiency
Hemoglobin	Normal	Normal	Reduced	Normal
MCV	Normal	Normal to reduced	Reduced	Reduced
Serum iron (SI)	Normal	Reduced	Reduced	Normal
Serum ferritin	Reduced	Reduced	Reduced	Normal to elevated
TIBC	Normal	Increased	Increased	Increased
sTfR	Normal	Increased	Increased	Increased
CHr or Ret-He	Normal	Decreased	Decreased	Decreased
Hepcidin	Reduced	Reduced	Reduced	Elevated

MCV: mean corpuscular volume; TIBC: total iron-binding capacity; sTfR: soluble transferrin receptor; CHr or Ret-He: reticulocyte hemoglobin content.

Table 1.
Classification of the iron states and associated laboratory findings.

Iron deficiency affects a variety of physiological functions [5]. Iron deficiency refers to the reduction of iron stores that precedes overt iron deficiency anemia or persists without progression. Iron deficiency anemia is a more severe condition in which low levels of iron are associated with anemia and the presence of microcytic hypochromic red cells [13].

Serum ferritin represents a small fraction of the body's ferritin pool, but the concentration of ferritin reflects the amount of iron stores. Once iron stores are depleted, the first stage of iron deficiency (ID) is reached, namely iron depletion, but there are no erythropoietic consequences yet [15].

The iron supply provided by the transport iron compartment is mainly for red blood cell (RBC) production because the demand of iron for erythropoiesis is much larger than other tissues. The second stage of ID, namely iron-deficiency erythropoiesis, occurs without showing a notable decrease in hemoglobin concentration, when the supply can no longer be met. Laboratory parameters providing information about the adequacy of iron supply are transferrin saturation (TSAT) and the concentrations of erythrocyte protoporphyrin (EP), and soluble transferrin receptor (sTfR). The percentage of binding sites on all transferrin molecules occupied with iron molecules is represented by TSAT, and is calculated as the ratio of serum iron to transferrin or serum iron to total iron-binding capacity (TIBC) [15].

Impairment of the delivery of iron to erythroid is indicated by iron-restricted erythropoiesis, no matter how replete the stores. In cases of anemia of chronic inflammation, stores may be normal or even increased because of iron sequestration, which is observed in patients with autoimmune disorders, infections, and chronic kidney diseases [13]. Common indicator considerations require biological confounding caused by the inflammation. Inflammation is a highly complex biological process, confounding the interpretation of iron status indicators, especially serum ferritin concentration because it increases in response to inflammation as well as to increased iron stores [15].

In uncomplicated IDA (without inflammation response), there is a reduction in iron stores, transport iron, and functional iron. Transferrin production is upregulated to increase iron transport, as soon as the iron supply to erythropoiesis becomes insufficient. Upregulation of transferrin receptor production happens to facilitate iron delivery to cells increasing sTfR, and zinc protoporphyrin (ZPP) is produced instead of heme resulting in an increase of erythrocyte protoporphyrin (EP). Serum ferritin and Hb concentration are important indicators in uncomplicated IDA [15].

The functional iron deficiency is a state of iron-poor erythropoiesis in which there is an insufficient mobilization of iron from stores in the presence of increased demands, as is observed after treatment with erythropoiesis-stimulating agents [13].

3.3 Diagnosis and laboratory findings

A detailed history (anamnesis) of the patient and physical examination is crucial in the diagnosis of all diseases in medical science. A study has shown that a detailed history can diagnose anemia with a sensitivity of 71% and specificity of 79% [16]. Particularly, prenatal period, times of starting breastmilk and solid foods, bleeding history and nutrition should be considered in detail, also signs other systemic diseases and anemia that may accompany [1].

A hemoglobin (Hb) value 5 percentile below the normal hemoglobin value specified for that age or reduced erythrocyte count in healthy individuals is defined as anemia. Anemia should be defined by paying attention to the lower limit of the normal value for different age groups and gender [1]. Hemoglobin concentration is the key indicator for a functionally important iron deficit, specifically iron deficiency anemia (IDA). The hematocrit does not reveal any additional information other than hemoglobin [15]. Based on the size of RBC, hematologists categorize the anemia as macrocytic, normocytic, or microcytic [12].

Anemia in children has a broad differential diagnosis, but it narrows once the anemia is classified further as microcytic. The most common causes of this in children are iron deficiency and thalassemia minor. Microcytosis also results from lead poisoning, chronic diseases (e.g., inflammation, infection, etc.), sideroblastic anemia, and other rare conditions [14].

Reduction in MCV and MCH (mean corpuscular hemoglobin) in a CBC result is a manifestation of reduced hemoglobin in erythrocytes. The erythrocytes are paler and smaller than normal on the peripheral blood smear, (microcytic and hypochromic). MCV and MCH are parallel to each other; meaning erythrocytes may be microcytic and hypochromic at the same time. An MCH below 27 pg. is considered low. The normal value of MCV ranges between 80 and 99 fL, but in children, normal values differ according to age. Laboratory findings in iron deficiency are shown in **Table** 2 below [1]. The data from our study (**Table 3**) shows significant differences in hematologic parameters between the β-thalassemia minor and IDA groups. The higher RBC increase in the IDA group compared to the β-thalassemia minor (BTMi) group was probably related to the administration of iron therapy in children with IDA [17].

Differential diagnosis of microcytic hypochromic anemia is very important to consider because the interpretation of its' peripheral blood can be found in iron

Complete blood count:
RDW >14
RBC: low
Hb, Hct: low according to age and gender
MCV: low according to age and gender
When specifying the lower limit of MCV: 70 + age (for >10 years)
(if MCV is <72, generally abnormal)
Upper limit of MCV: 84 + age x 0.6 (for >6 months)
(if MCV is >98, always abnormal)
MCH <27 pg.
MCHC <30%
Thrombocytosis
Rarely: thrombocytopenia, leukopenia

Peripheral blood smear:
Hypochromic
Microcytosis
Anisochromic
Anisocytosis
Pencil cells
Rarely: basophilic stippling, target cells, hyper segmented neutrophils
Serum ferritin <12 ng/mL
*Serum iron <30 mcg/dL
*TIBC >480 mcg/dL
Transferrin saturation (SI/TIBC × 100%) <16%
Mentzer index (MCV/RBC) <13

May change by age, gender, and other factors. Should be evaluated together.

Table 2.
Iron deficiency laboratory findings.

Parameter	BTMi (n: 159)		IDA (n: 64)	
	Range	Mean ± SD	Range	Mean ± SD
Hb (g/dL)	4.49–14	8.53 ± 1.62*	5.07–16.3	10.96 ± 2.13*
Hct (%)	13.6–43.8*	27.57 ± 5.06	14.7–51.6*	34.15 ± 6.89
RBC (×10⁶/µL)	1.9–6.77*	3.86 ± 0.81	2.06–6.05*	4.46 ± 8.86
MCV (fL)	55.0–99.3	71.95 ± 6.76*	63.4–90.7	76.48 ± 4.85*
MCH (pg)	16.6–31.7	22.34 ± 3.02*	20.0–28.7	24.59 ± 2.11*
MCHC	26.5–34.9	30.96 ± 1.86*	27.6–34.9	32.20 ± 1.45*
RDW-CV (%)	8.3–34*	20.15 ± 4.77	10.5–30.9*	14.6 ± 3.28

Note: Hb: hemoglobin; RBC: red blood cell; MCV: mean corpuscular volume; MCH: mean corpuscular hemoglobin; MCHC: mean corpuscular hemoglobin concentration; RDW-CV: red cell distribution width-coefficient of variation. *Significant, $p < 0.001$.

Table 3.
Hematological parameters of the group of β-thalassemia minor (BTMi) and iron deficiency anemia.

deficiency anemia and β-thalassemia trait. Iron deficiency and β-thalassemia minor are best differentiated using serum ferritin level, serum iron level, total iron-binding capacity, transferrin saturation, and Hb A2 level, along with a complete blood count (CBC) and examination of peripheral blood film [18]. Carriers of β-thalassemia are usually clinically asymptomatic. However, they have characteristics of the CBCs with mean corpuscular volume (MCV) less than 80 fL and mean corpuscular hemoglobin (MCH) less than 27 p. [19].

Anemia evaluation can be done by an array of tests, but there is no single "best" test to diagnose iron deficiency, with or without anemia. The "gold standard" for identifying iron deficiency is bone marrow biopsy with Prussian blue staining. Since, bone marrow aspiration is an invasive procedure, indirect assays are used for routine use. The laboratory tests that may be used to support and consider the diagnosis of iron deficiency are complete blood count (CBC), peripheral blood smear, reticulocyte, iron profile (SI, TIBC, and transferrin saturation index), sTfR level, and biochemical tests based on iron metabolism (e.g., zinc protoporphyrin-ZPP, serum ferritin concentration) [14]. In CBC, if anemia is present, it should be primarily checked if hemoglobin and hematocrit values are normal for age and gender. In infants younger than 6 months, lower values are observed because of physiological anemia, but hemoglobin levels are not expected to be lower than 9 g/dL in physiological anemia in term infants if there is no other accompanying factor [1].

4. Conclusion

Iron has a role in various essential physiological functions, such as oxygen transport, gene regulation, DNA synthesis, DNA repair, and brain function. Depletion of and inability to use iron disturbs these pathways and causes multiple morbidities. Iron deficiency anemia (IDA) is a well-known complication, but iron deficiency alone may cause negative impacts on the health risk of pediatric patients.

The high prevalence of iron deficiency anemia in developing countries most often is attributed to nutritional deficiencies worsened by chronic blood loss due to parasitic infections and malaria. The differential diagnosis for anemia in children is broad, but it narrows once the anemia is classified further as microcytic. Iron deficiency and thalassemia minor are the most common causes of microcytic anemia in children.

An array of tests can be used for evaluating anemia, but there is no single "best" test to diagnose iron deficiency, with or without anemia. The "gold standard" for identifying iron deficiency is bone marrow biopsy with Prussian blue staining. The laboratory tests that may be used to support and consider the diagnosis of iron deficiency are complete blood count (CBC), peripheral blood smear, reticulocyte, iron profile (SI, TIBC, and transferrin saturation index), sTfR level, and biochemical tests are based on iron metabolism.

Conflict of interest

The authors declare no conflict of interest.

Author details

Yulia Nadar Indrasari[1]*, Siti Nurul Hapsari[2] and Muhamad Robiul Fuadi[1]

1 Faculty of Medicine, Hematology Division, Department of Clinical Pathology, Airlangga University, Dr. Soetomo General Academic Hospital, Surabaya, Indonesia

2 Faculty of Medicine, Department of Clinical Pathology, Airlangga University, Surabaya, Indonesia

*Address all correspondence to: ynadar.indrasari82@gmail.com

IntechOpen

References

[1] Özdemir N. Iron deficiency anemia from diagnosis to treatment in children. Turkish Archives of Pediatrics. 2015;**50**(1):11-19

[2] WHO. The global prevalence of anaemia in 2011. In: Peña-Rosas JP, Rogers L, Stevens G, editors. The Global Prevalence of Anaemia in 2011. 1st ed. Geneva: WHO press; 2015. pp. 1-48. Available from: https://apps.who.int?iris/hendle/10665/177094

[3] Wu AC, Lesperance L, Bernstein H. Screening for Iron deficiency the early introduction of whole cow milk. Pediatrics in Review. 2011;**23**(5):171-178

[4] Domellöf M. Iron requirements, absorption and metabolism in infancy and childhood. Current Opinion in Clinical Nutrition and Metabolic Care. 2007;**10**(3):329-335

[5] Tong S, Vichinsky E. Iron deficiency: Implications before. Anemia. 2021;**42**:1

[6] Lonnerdal B, Georgieff M, Hernell O. Developmental physiology of Iron absorption, homeostasis and metabolism in the healthy term infant. The Journal of Pediatrics. 2015;**167**(40):S8-S14

[7] Lönnerdal B, Kelleher SL. Iron metabolism in infants and children. Food and Nutrition Bulletin. 2007;**28**(Suppl. 4):S491-S499

[8] Domellöf M. Iron requirements in infancy. Annals of Nutrition & Metabolism. 2011;**59**(1):59-63

[9] Eussen S, Alles M, Uijterschout L, Brus F, Van Der Horst-Graat J. Iron intake and status of children aged 6-36 months in Europe: A systematic review. Annals of Nutrition & Metabolism. 2015;**66**(2-3):80-92

[10] Wang Y, Wu Y, Li T, Wang X, Zhu C. Iron metabolism and brain development in premature infants. Frontiers in Physiology. 2019;**10**(APR):1-13

[11] Gunshin H, Fujiwara Y, Custodio AO, DiRenzo C, Robine S, Andrews NC. Slc11a2 is required for intestinal iron absorption and erythropoiesis but dispensable in placenta and liver. The Journal of Clinical Investigation. 2005;**115**(5):1258-1266

[12] Wu AC, Lesperance L, Bernstein H. Screening for iron deficiency. Pediatrics in Review. 2002;**23**(5):171-178

[13] Camaschella C. Iron-deficiency Anemia. The New England Journal of Medicine. 2015;**372**(19):1832-1843. DOI: 10.1056/NEJMra1401038

[14] Wu AC, Lesperance L, Bernstein H. Screening for Iron deficiency. Pediatrics in Review. 2002;**23**(5):171 LP-171178. Available from: http://pedsinreview.aappublications.org/content/23/5/171.abstract

[15] Pfeiffer CM, Looker AC. Laboratory methodologies for indicators of iron status: Strengths, limitations, and analytical challenges. The American Journal of Clinical Nutrition. 2017;**106**(Suppl. 6):1606S-1614S

[16] Boutry M, Needlman R. Use of diet history in the screening of iron deficiency. Pediatrics. 1996;**98**(6 Pt 1): 1138-1142

[17] Indrasari YN, Hernaningsih Y, Fitriah M, Hajat A, Ugrasena IDG. Reliability of different RBC indices and formulas in the discrimination of β-thalassemia minor and Iron deficiency Anemia in Surabaya, Indonesia. Indian

Journal of Forensic Medicine and
Toxicology. 2021;**15**(1):984-989

[18] Keohane EM, Otto CN, Walenga JM.
In: Keohane EM, Otto CN, Walenga JM,
editors. Rodak's Hematology: Clinical
Principles and Applications. Sixth ed. St.
Louis, Missouri: Elsevier; 2020.
pp. 424-442

[19] Bordbar E, Taghipour M,
Zucconi BE. Reliability of different rbc
indices and formulas in discriminating
between β-thalassemia minor and other
microcytic hypochromic cases.
Mediterranean Journal of Hematology
and Infectious Diseases. 2015;7(1):1-7

Chapter 8

FERALGINE™ a New Oral iron Compound

Valentina Talarico, Laura Giancotti, Giuseppe Antonio Mazza, Santina Marrazzo, Roberto Miniero and Marco Bertini

Abstract

Management of iron deficiency (ID) and iron deficiency anemia (IDA) is primarily focused to remove, when possible, the underlying cause of ID; subsequently its treatment is primary focused on iron stores repletion. Ferrous sulphate (FS) remains the mainstay of treatment and it is recommended as the first-line treatment of ID and IDA in children as in adults by all guidelines of scientific societies. However the effectiveness of FS is largely compromised by increased adverse effects, poor compliance and discontinuation of treatment. A new oral iron source named FERALGINE™ (FBC-A) has been recently developed. This new molecule is a patented co-processed one-to-one ratio compound between Ferrous Bysglicinate Chelate (FBC) and Sodium Alginate (AA), obtained by using a spray drying technology. The data presented in this short review highlight the efficacy and safety of the treatment with FBC-A and support its use in adult patients with IDA. Furthermore the present review also provides preliminary evidence to suggest FBC-A as first-line treatment for ID/IDA in patients with celiac disease (CD) or inflammatory bowel diseases (IBD).

Keywords: Iron deficiency, Iron deficiency anemia, oral treatment, compliance

1. Introduction

Iron deficiency (ID) is the most common nutritional deficiency worldwide, heavily concentrated in several regions including Asia, Latin America and Africa, where it may affect up to 60% of the entire population. In countries with high development rate prevalence of iron deficiency anemia (IDA) is estimated at 9% and accounts approximately for 50% of all anemia cases, representing a frequent medical condition encountered in clinical practice by general practitioners, pediatricians and several other specialists. In these countries the prevalence of ID/IDA is higher in pediatric age, especially in two life phases: one that occurs between the first and third year of life (2.3–15%) and another in adolescence (3.5–13% in males, 11–33% in females). In adults its prevalence is less than 1% in men <50 years of age, 2–4% in men >50 years of age, 9–20% in menstruating teenagers and young women, and 5 to 7% in postmenopausal women. In people older than 65 years its prevalence is 12%. World Health Organization (WHO) data show that ID/IDA in pregnancy is a significant problem throughout the world with a prevalence ranging from an average of 14% of pregnant women in industrialized countries to an average of 56% (range 35–75%) in developing countries [1–5].

The most frequent etiologic factors in ID/IDA are decrease iron intake, impaired iron absorption, increase iron loss and increased iron requirements. In adult females increased menstrual flows and reduced iron absorption, as occurs in celiac disease (CD), are the most common mechanisms of ID/IDA. Vegetarians, especially vegans, obese individuals, blood donors and competitive endurance athletes represent populations at risk of ID/IDA. In children increased daily requirement and CD are the leading causes of ID/IDA [1–5].

Iron (Fe) is an essential trace mineral naturally present in many foods, but it is also available as a dietary supplement. In the typical western diet iron is mainly (85–90%) present as inorganic form Fe^{2+} (ferrous) and Fe^{3+} (ferric), the remaining amount as heme form. The diet contains up to 20 mg of daily iron intake of which 1–2 mg are absorbed. Fe^{3+} is less bioavailable as it has to be converted into Fe^{2+} in order to be absorbed. The absorption of Fe^{2+} primarily occurs in the proximal duodenum, at the brush border of the mucosa cells, through a membrane transport protein called Divalent Metal Transporter 1 (DMT1). This process is regulated by the cytochrome B (DCYTB), a ferric reductase located on the apical membrane of duodenal enterocytes. Otherwise heme-iron is absorbed in the same bowel district, but separately from DMT1 and more efficiently than inorganic Fe [1–5].

Management of ID/IDA is primarily aimed at removing, whenever possible, the underlying cause of ID. Subsequently iron replacement is always indicated to replete iron stores. The use of oral iron formulations is the current standard treatment; however in certain situations discussion remain open if the intravenous iron might be a more suitable modality for iron supplementation [1–5].

Ferrous sulphate (FS) remains the mainstay of treatment since it was first introduced in 1832 by the French physician Pierre Blaud (*Blaud's pills*) and it is recommended as the first-line treatment of ID/IDA in children and in adults by all guidelines of scientific societies. Two other bivalent iron preparations are primary suggested: ferrous gluconate and ferrous fumarate. None of these compounds seem to be better than the others [1–5].

The optimal oral iron dose of FS is yet to be established. Traditionally recommended dose in children is 2–6 mg/kg/day in term of elemental iron [5]. In adolescents and adults recommended dose is 50–200 mg once daily or in divide doses [1–4]. The optimum frequency of oral supplementation is still uncertain. It has been demonstrated that one day treatment saturates the intestinal absorption processes. According to more recent data of the literature the administration of iron on every other day might be equal or more effective than daily doses with less side effects [2, 4, 6]. When therapy is fully effective the anticipated increase in hemoglobin levels occur after 2 to 3 weeks (increase by 1–2 g/dl within 1 month) of iron treatment, and reaches normal levels by 2–3 months. When the hemoglobin levels have been corrected, treatment should be continued for 3–4 months in order to completely fill the body's iron stores [1–5].

Ferrous sulphate absorption ranges between 5–28% at the fastest. During oral iron therapy non-absorbed iron is potentially toxic for the gastrointestinal mucosa due to its oxidative properties leading to occurrence of gastro-intestinal adverse events. Nausea, vomiting, diarrhea or constipation, epigastric discomfort and colicky pain often represent a limit for patient domiciliary compliance decreasing the adherence to protocols with consequent failure of therapy efficacy. Actually many patients (20–70%)

experience some type of gastrointestinal discomfort during oral iron salts intake, jeopardizing the prolonged (several months) planned treatment [1–7]: up to 40% of patients may self-discontinue the medication without discussing with medical doctor [8]. It is not surprising that effectiveness of oral iron is largely compromised by lack of absorption, poor compliance, increased adverse effects and discontinuation of treatment.

In order to improve tolerability several formulations with Fe2+ or Fe3+ have been proposed by pharmaceutical laboratories during the last decades. Generally these oral iron compounds are better tolerated than FS but may be less effective in iron replacement, ferrous compounds remaining anyway more absorbed than trivalent ones. However, often these preparations have the common drawback of being less effective in malabsorptive disorders [1–7]. Clinical studies concerning these new compounds remain limited while rigorously randomized designed clinicaltrials are often lacking.

Iron amino acid chelates represent a source of iron which has proven to be highly bioavailable with decreased extent gastrointestinal adverse effects when compared to FS [9]. Ferrous Bisglycinate Chelate (FBC) is the most studied compound among these new formulations. In FBC one molecule of ferrous iron is chelated by two molecules of glycine resulting in two heterocyclic rings. Several clinical trials showed clinical bioequivalence between this source of iron at low dosage and FS at standard doses with ratio 1 to 4 [10–12]. Unfortunately, during treatment with FBC, albeit very rare, some gastro-intestinal adverse effects can occur. Another limit of FBC could be represented by "the iron taste" of the preparation that might worsen patient compliance.

To ameliorate its bioavailability, taste and tolerability, a new oral iron source named FERALGINE™ (FBC-A) has been recently developed [13, 14]. This new molecule is a patented co-processed one-to-one ratio compound between FBC and Sodium Alginate (AA), obtainedby using a spray drying technology [15, 16]. AA as well as FBC has been recognized as GRAS by FDA [17].

Sodium Alginate has usually been used as "gastro-protection". It is a non-toxic, biocompatible, biodegradable polymer, which belongs to the polysaccharides naturally present in seaweed. The contact between AA and the acidic environment in the stomach leads to the formation of a gel layer that hasa protective effect on the mucosa membranes of the stomach and esophagus. For this reason AA is an ingredient of many medications (antacids) commonly used in the treatment of heartburn and reflux diseases [18, 19]. Mucoadhesive microsphere of AA have recently demonstrated to be a unique "carrier system" for many pharmaceuticals preparation increasing not only oral iron but also other drugs bioavailability such as metformin, amoxicillin, furosemide, ibuprofen, insulin, acyclovir, captopril, glipizide, dicumarol. This "carrier system" results really in a promising and cost-effective method drug delivery system to improve oral bioavailability and to reduce gastrointestinal drugs side effects [20–22]. Spray drying is one of the most powerful technological process for the pharmaceutical industry << being an ideal process where the moisture content, bulk density and morphology end-product must comply with precise quality standards regarding particle size distribution, residual moisture content, bulk density and morphology>> [15, 16, 23–25]. (**Figure 1**) Spray drying technology comes of age during World War II, with the sudden need to reduce the transport weight of foods and other materials. This technique enables the transformation of feed from a fluid state into dried particulate form by spraying the feed into a hot drying medium. It is a continuous particle processing drying operation. As well described by Gervasi et al. < <spray drying process mainly involves five steps: 1) Concentration: feedstock is normally concentrated prior to introduction into the spray dryer; 2) Atomization: the atomization stage creates the optimum condition for evaporation to a dried product

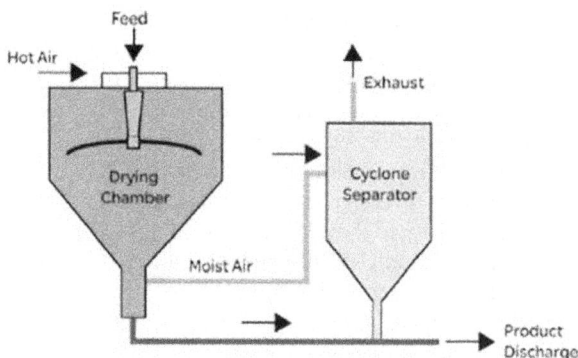

Figure 1.
Spray drying system.

having the desired characteristic; 3) Droplet-air contact: in the chamber, atomized liquids brought into contact with hot gas, resulting in evaporation of 95% of the water contained in the droplets in a matter of a few minutes; 4) Droplet drying: moisture evaporation takes place in two stages: during the first stage, there is sufficient moisture in the drop to replace the liquid evaporated at the surface and evaporation takes place at a relatively constant rate and the second stage begins when there is no longer enough moisture to maintain saturated conditions at the droplet surface, causing a dried shell to form at the surface. Evaporation then depends on the diffusion of moisture through the shell, which is increasing in thickness; 5) Separation: cyclones, bag filter, and electrostatic precipitators may be used for the final separation stage>> [14].

In FBC-A, since FBC and AA are present in 1 to 1 ratio, every little particle of the powder has the same morphology and quantity of the two different co-processed substances. The new "co-processed compound", obtained by spray drying technology, confers to the iron powder an increased and uniform superficial area and, consequently, quick and more extensive iron absorption together with an increased gastrointestinal protection (**Figure 2**). In the same time, the uniform presence of AA in this

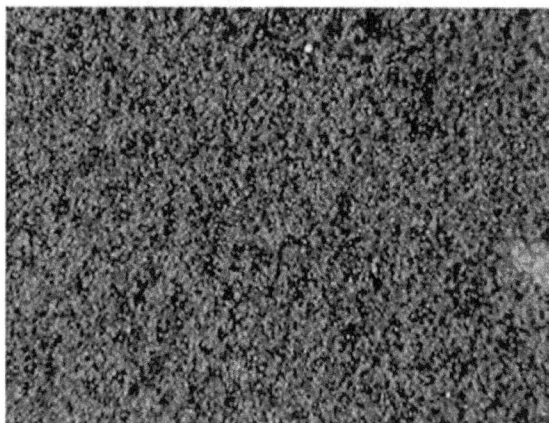

Figure 2.
FBC-A powder (picture making by stereomicroscopy Wild Heerbrugg Makroskop M420 linked to an OPTIKAM MICROSCOPY DIGITAL USB CAMERA) [14].

product, allows theFBC to be released more constantly and slowly when confronted to FBC alone and let the DMT1 receptors to better uptake iron. DMT1 receptor saturation could be a limit in oral iron bioavailability that could be exceeded by FBC-A. In fact, slowly availability of iron by FBC-A administration could result in DMT-1 unsaturation with the consequence of increasing iron bioavailability [13, 14, 23–25].

2. Clinical studies

Preliminary clinical trials confirm the bioavailable of FBC-A whiles the "dispert effect" of AA on FBC accounts for the good tolerability at gastro-enteric levels. FBC-A also improves iron taste when compared to FBC alone, increasing patient's compliance.

Ame et al. studied 12 patients (9 women and 3 men with medium age of 63.83 ± 20.94 years) affected by IDA (4 patients present multifactorial anemia, 2 hypermenorrhea-related anemia, 2 cancer-related anemia, 2 increased-iron loss anemia, 1 post transplantation anemia and1 hypo-regenerative anemia), enrolled in an open prospective uncontrolled pivotal clinical trial. Mean hemoglobin (Hb) levels at the beginning of study (T0) were 10.49 g/dl (range 7.8–11.9 g/dl), mean serum iron values were 27,9 µg/dL (range 13–39 µg/dL) and mean serum ferritin (SF) values were 26 µg/ml (range 4–89 µg/ml). The patients presented history of chronic fatigue and/or asthenia at enrolment. All patients received FBC-A (30 mg of elemental iron) once a day for a period lasting from 35 days to 60 days (mean 46.25 days). At the end of treatment (T1) mean Hb values were 11.6 g/dl (range 8.9–13.9 g/dl), mean serum iron values were 48.9 mg/ld. (range 34–68 mg/dl) and mean SF values were 35 µg/ml (range 9–94 µg/ml), (p < 0.0001) (**Figure 3**). No FBC-A adverse events or therapy interruption were reported during the trial. Significant as a small quantity of elemental iron (30 mg daily) has been able to increase the Hb as required by international guidelines (1 g/dl.Hb/month) confirming the high bioavailability ofthis new compound. Subject performances ameliorated significantly in all patients [26].

Celiac disease (CD) is an immunologically-mediated disorder characterized by duodenal mucosa villi atrophy. As iron is primarily absorbed at duodenum level, iron absorption is reduced in celiac patients. In fact, the most frequent extra-intestinal manifestation of CD is IDA, with a prevalence between 12 and 82% in patients with new CD diagnosis. Absorption of FS and other iron formulationsis limited in patients with undiagnosed and active CD. Iron supplementation generally results less effective in these patients leading to a form of refractory IDA. The effectiveness of iron administration may be reduced also during the first months of gluten-free diet (GFD), when

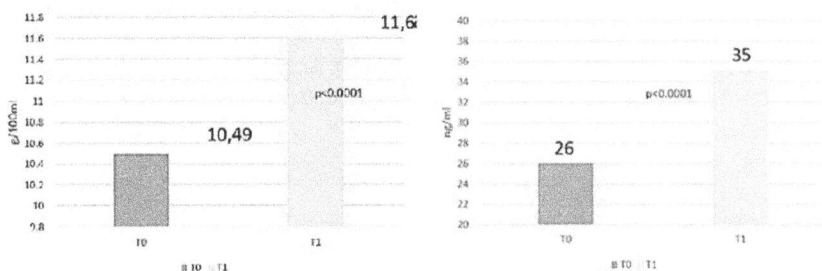

Figure 3.
Increase in mean Hb and SF levels from time T0 to T1 after administration of FBC-A [26].

the mucosa healing is yet on-going. Furthermore a poor iron compoundstolerabilityis particularly frequent in patients with CD, decreasing patient's compliance. These events increase the risk of making every kind of oral iron treatment unhelpful [27, 28].

The oral iron absorption test (OIAT) is an old screening test to assess iron absorption. Forgotten for many years it has been recently re-evaluated in clinical practice [29, 30]. The test consists of measuring plasma iron increase in the next hours after a single dose of an oral iron preparation. It has been demonstrated that this increment reflects the real capacity of iron absorption from the gastrointestinal tract [29, 30]. In a previous study we tested FBC by OIAT in pediatric patients affected by overt CD (disease at diagnosis before starting GFD) or on GFD, showing as FBC was well absorbed in spite of duodenal mucosal lesions [31].

More recently Giancotti et al. studied 26 patients with IDA of which 14 were also affected by overt CD (mean age: 32.28 years) and 12 were not affected (mean age: 33.58 years) [32]. The demographic and laboratory baseline parameters of the patients are summarized in **Table 1.**

An OIAT was performed in each patient by administrating FBC-A (60 milligrams of elemental iron). Serum iron was evaluated at baseline (T0) and after 2 h (T1) from the iron ingestion. The OIAT was well tolerated in all patients. There was a clear improvement in iron serum in all patients (T0 = 31.30 µg/dL vs. T1 = 105.3 µg/dL # p < 0.0001). The relationship between the severity of IDA and the absorption of iron showed that patients with severe anemia (Hb < 10 g/dL) had an higher increase in serum iron after the OIAT (about nine times) compared to patients with mild/moderate forms of anemia (TO 12.00 µg/dL and 35.76 µg/dL *vs* T1 109.20 µg/dL and 104.66 µg/dL in severe anemia and mild/moderate anemia, respectively). Surprisingly, an equivalent improvement in serum iron occurred in the two groups of patients (IDA plus CD and IDA without CD): T0 = 28.21 µg/dl *vs*. T1 = 94.14 µg/dl, (p = 0.004) in the first group (Group A) and T0 = 34.91 µg/dl *vs*. T1 = 118.83 µg/dl (p = 0.0003), in the other group (Group B) respectively as shown in **Figure 4** [32]. All the 26 patients were compliant to the treatment with FBC-A (60 milligrams of elemental iron once a day) and continued it until normalization of Hb levels and SF that occurred after 3 and 5 months. Response to treatment was monitored with periodical evaluation of Hb, serum iron, transferrin saturation and SF (unpublished personal data).

These results clearly demonstrated that FBC-A is well tolerated and well absorbed, not only in anemic non-celiac patients but also in patients with overt CD. Furthermore as it is widely assumed that side effects limit compliance to iron oral treatment, these results confirm that FBC-A is very promising for treatment of iron deficiency also in patients affected by CD.

Talarico et al. referred a 22-year-old girl with IDA resistant to FS therapy who was finally diagnosed as a celiac patient with multiple duodenal biopsies. A strictly GFD was then prescribed and in order to evaluate the absorption of a different iron compound, an OIAT was performed with FBC-A at the dosage of 30 mg of elemental iron. The results confirmed a good iron absorption as the serum iron increased from 27 µg/dl at

	Males	Females	Hb ± SD (g/dl)	Serum ferritin±SD (ng/dl)	Serum iron ±SD µg/dL)
Celiac-IDA	2	12	11.07 ± 1.04	10.44 ± 15.9	28.21 ± 14.9
Non-celiac IDA	0	12	10.80 ± 0.9	12.30 ± 13.8	34.91 ± 23.2

Table 1.
The demographic and laboratory baseline parameters of the patients [32].

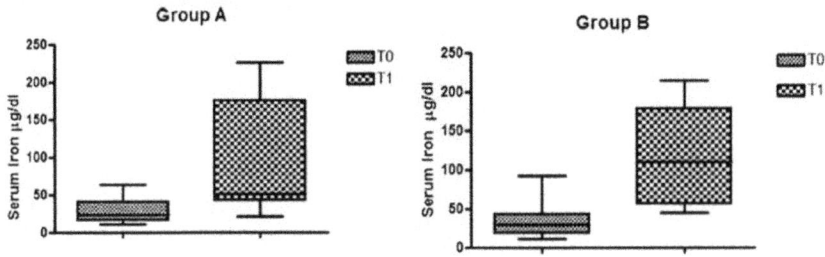

Figure 4.
Change in serum iron levels in patient with IDA and CD (group A) and without CD (group B) after OIAT [32].

baseline (T0) to 93.2 µg/dl after two hours (T1). Then, treatment with FBC-A at a dosage of 30 mg/day was promptly started. A general status recovery was observed in a few days. After 3 months Hb reached 11.5 g/dl; after 6 months iron stores were replaced (SF 50 ng/dl). CD serology normalized after one year, with complete resolution of anemia and gastrointestinal symptoms [33]. This case report confirms the efficacy of FBC-A in patients with overt CD, being already well absorbed during the first months of GFD.

Rondinelli et al. performed an OIAT study after oral ingestion of FBC-A (60 mg of elemental iron) on 14 patients (8 females and 6 males) with IDA (mean age 55.28 ± 8.17 years). The mean Hb level was 9.7 g/dl (range 8.7–10.7g/dl) and mean SF level was 8.23 ng/ml (4–12 ng/ml). At baseline time (T0) mean plasma iron was 11.21 µg/dl ± 10.66. Two hours after taking FBC-Ain fasting condition (T2) mean plasma iron increased significantly to 111.00 µg/dl ± 51.56 (p < 0.00001) (**Figure 5**) [34].

The increase of serum iron was greater than that observed after administration of bivalent iron by others authors (**Figure 6**).

Crohn's disease and ulcerative colitis are the two expressions of the Inflammatory Bowel Disease (IBD), a complex of immunologically mediated diseases due to a dysregulated immune response to commensal flora in a genetically susceptible host. Among IBD patients, IDA is more frequent than in general population, due to multifactorial reasons including chronically inflammation, blood loss and low iron absorption. Goodhand et al. showed as children (88%) and adolescents (83%) were more often

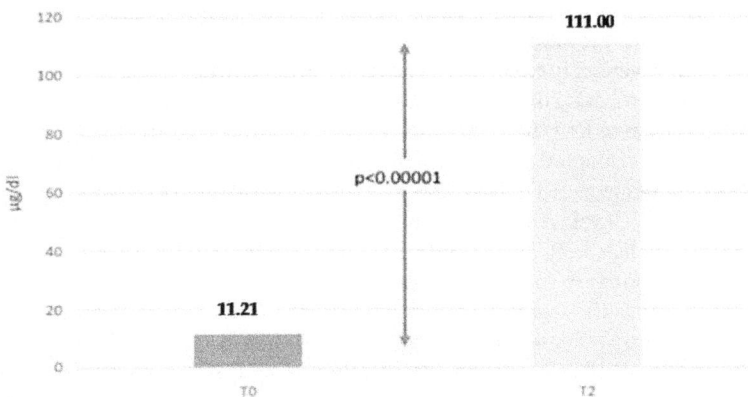

Figure 5.
OIAT: plasma iron before (T0) and after 2 hours (T2) from 60 mg of FBC-A [34].

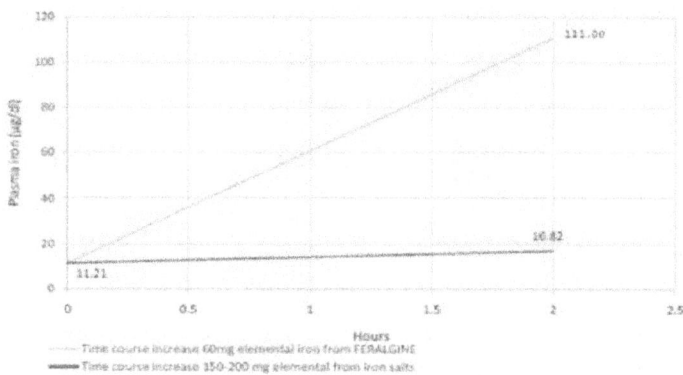

Figure 6.
Expected theoretical plasma iron level after 2 hours from 200 mg of elemental iron salts administration (blue lines) versus experimental FBC-A 60 mg elemental iron oral administration (yellow line) [34].

iron-deficient than adults (55%). The efficacy of oral iron treatment might be hampered by a significant incidence of gastrointestinal side effects and the exacerbation of an existing colitis, that limit adherence to therapy in near half of patients [35]. It has been shown that in adults the adherence to therapy lowers to 10–42% within 2 months. Consequently treatment with oral iron results in failure to control anemia in 2 out of 3 IBD patients as reported by Lugg et al. [36] According to the last European Crohn's and Colitis Organization [ECCO] guidelines, oral iron therapy is recommended for patients with mild IDA and clinically inactive IBD, with no history of intolerance to oral iron; instead intravenous (IV) administration iron should be considered in patients with clinically active IBD with previous intolerance to oral iron [37]. However although adverse reactions to modern IV iron formulations are rarely severe, given the potentially fatal side effects this treatment is generally only performed in a hospital setting where ready access to resuscitation equipments is available. Furthermore IV treatment has a significant incremental cost respect to oral iron treatment [37]. A recent review by Foteinogiannopoulou et al. on 1394 Greek patients with IBD showed that among those who received iron intravenously (393) about 89.3% responded, in contrast to only 54.2% of those who received iron orally (142). Furthermore 31% of patients that received iron orally presented with adverse events, mostly gastrointestinal symptoms that eventually led to cessation ofthe treatment. On the other hand, about 7.9% that received iron intravenously experienced adverse events during infusion. The availability of a well absorbed oral compound with good tolerability would be of great benefit for these patients, avoiding the need of intravenous treatment [38].

Vernero et al. performed a study with FBC-A in patients affected by IBD. They enrolled 52 patients with IBD, mean age 47 years (range 15–86 years) presenting with IDA. Mean Hb level was 11 g/dl (range 10.72–11.47 g/dl) and mean SF level was 21.4 ng/ml (range 3.09–39.7 ng/ml). Patients received FBC-A 30 mg once a day for 3 months. Response to therapy was monitored by periodical evaluation of Hb and SF. At the end of the study Hb mean value was 12.2 g/dl (range 11.6–12.52 g/dl) (p = 0,0001). Mean SF increased to 74 ng/ml (range 14.4–133.7 ng/ml), a difference near to the statistical significance (p = 0,07). Regarding FBC-Atolerability, 90% of patients reported good tolerance to treatment, while 10% of them experienced dyspepsia and worsening of diarrhea. Only 6% of patients suspended oral iron supplementation due to gastroenteric intolerance (adherence rate 94%) [39].

Sprecacenere et al. evaluated the impact of a 4 weeks course of FBC-A (30 mg/day) on a cohort of 52 blood donors presenting with IDA (mean Hb level11.8 ± 0.7 gr/dl and mean SF level 14.8 ± 9.8 ng/ml). After treatment mean Hb level increased to 13.3 ± 1 gr/dL (p = 0,007) and mean SF level to 21.8 ± 13.5 ng/ml (p = 0,02) [40].

Iron requirements drastically increase during pregnancy to accommodate an expandingred cell volume, growing fetus and placenta. The global prevalence of anemia in pregnancy is estimated to be approximately 38–42%. ID may be present in up to 80% of women in the second/third trimester of pregnancy and postpartum, if not supplemented with iron. Oral iron treatment is still considered a frontline therapy for IDA in pregnancy but gastrointestinal side effects are common leading to a high ratio of non-adherence to treatment [41]. Bertini et al. described 21 pregnant women with moderate IDA (Hb < 11 g/dl and SF < 25 ng/ml at time T0) supplemented with 30 mg/day of elemental iron in the form of FBC-A and evaluated at 15 days (T1) and 30 days (T2) of treatment. Hb and SF levels increased in all the enrolled women at T1 and T2 interval (Hb T1 vs. T0 p < 0.004; T2 vs.T0 p < 0.00001 and T2 vs. T1 p < 0.003) (SF T1 vs. T0 p < 0.0001; T2 vs. T0 p < 0.00001; T2 vs. T1 p < 0.05). This study provided evidence that a small quantity of elemental iron (30 mg/day in form of FBC-A) administered by oral route is rapidly effective in restoring Hb and SF levels in pregnant women affected by moderate IDA, without gastrointestinal adverse effects [42].

3. Discussion

It is well documented that ID/IDA impacts negatively on the affected patients by disturbing multiple organs function possibly leading to a multitude of symptoms that compromise the general well-being, quality of life and individual performance. Successful management of ID/IDA primarily requires identification and treatment of the underlying cause(s) of the ID. After that replacement should be established for all patients with IDA/IDA with or without symptoms. Oral iron salts such as ferrous fumarate, ferrous gluconate, and ferrous sulfate have been the mainstay of oral iron supplementation as they are inexpensive, effective at restoring ironbalance, and have a good overall safety and tolerability profile. However, in several patients absorption of oral iron salts is inadequate, and poor tolerance results in reduced adherence to therapy. Some other compounds are nowavailable as alternative therapies. Usually they have a more tolerability profile over the traditional iron salts but often no studies comparing the clinical or cost effectiveness of these different oral iron products are available.This short review summarizes all clinical studies on FBC-A published so far.

Several studies demonstrated that OIAT is a good index for the evaluation of iron absorption and might a reliable test to investigate the bioavailability of the various iron formulations [29–31]. The studies of Giancotti et al. [32], Talarico et al. [33] and Rondinelli et al. [34] performing OIAT clearly show as the FBC-A is well absorbed and well tolerated. Furthermore these studies evidence as the compound is useful to restore Hb and SF levels confirming as combination between FBC and AA adds a very safe profile in terms of gastrointestinal adverse events and taste, improving patient's compliance.

Most of patients with GI has impaired iron absorption and develop ID/IDA. They often do not tolerate FS or other ironcompounds, leading to forms of refractory IDA resistant to treatment. Our review clearly show as FBC-A may be effective in CD [32, 33] and IBD [39] that represent frequent causes of refractory IDA. Then this new oral iron formulation appears promising in terms of safety and efficacy for these patients and might be suggested as first-line treatment. So far there are not studies regarding the use of

FBC-A in other rare forms of impaired iron aborption as those found in gastrectomized patients or in patients with achloridria. Parenteral iron treatment is often recommended in these conditions. We believe that FBC-A might play a role in their management. However future studies are advisable.

Our previous study clearly showed that FBC is well absorbed in pediatric patients with CD [31]. Two other studies from our group reported in this review confirmed the good absorption and tolerability also for FBC-A in celiac patients with overt CD and during the first months of GFD [32, 33]. In our opinion these results support the possibility that this preparation might be considered the treatment of choice for celiac patients with ID/IDA. Despite these clinical results the mechanism underlying the good absorption of FBC-A in CD remains unclear. Many studies have shown that aminoacid-chelated iron is better absorbed and better tolerated than FS and other types of inorganic iron [9–12]. In particular it has been demonstrated that low dosage of FBCis equal or better than standard dosage of FS in adults, in infants and in children [43–45]. The intestinal absorption of non-heme iron is mediated by DMT1 that is less expressed in CD as a consequence of the mucosa's lesions. This situation may explain why non-heme iron is less absorbed in this disease. Recent studies in pigs showed how FBC increases transcription of DMT1 and PepT1 genes, this latter coding for a heme-iron transporter. Therefore, the author suggested that FBC might be absorbed also as heme-iron, via the PepT1 [46]. Other studies revealed as FBC is absorbed intact by intestinal cells, highlighting the possibility that FBC absorption may occur also independently of the presence of DMT1 [47]. Moreover, latest findings of a slow and constant release of iron from FBC-A might explain the high bioavailability of iron, also in CD patients [25].

When to use IV versus oral iron administration in patients with IBD still represents an on-going topic of debate between clinicians [48]. According to the last ECCO consensus paper the usual treatment of IDA per os has relevant limitations in IBD patients, in factoral iron formulations use is restricted due to poor tolerability and patient compliance. The same consensus paper underlines as IV iron is more effective, shows a faster response, and is better tolerated than oral iron [37]. However although new IV iron formulations have proved to be bettertolerated and lead to a faster Hb rise than oral iron, there is still hesitancy among gastroenterologists to promote this administration due to its risk of hypersensitivity reactions [38]. The results obtained by Vernero et al. on IBD patients highlight as FBC-A is effective and well tolerated in subjects with intestinal mucosa lesions who generally are poor responders to oral iron treatment and have reduced compliance. If these data will be confirmed on larger cohorts, FBC-A might be considered a good candidate for the first-line treatment in patients with IBD and ID/IDA, avoiding IV approach. It would be interesting to compare FBC-A treatment versus IV formulations in large comparative trials.

Milman et al. showed as FBC 25 mg iron is as effective as FS 50 mg iron in the prophylaxis of ID/IDA during pregnancy in a randomized trial [43]. The paper of Bertini et al. confirmed the benefit of FBC-A 30 mg/die in pregnancy. Considering the good absorption and the low side effects, FBC-A might be an useful alternative to FS in pregnant women who often display poor compliance to oral iron preparations.

Oral iron is often poorly tolerated in patients with iron deficiency secondary to infection, inflammation, renal and malignant diseases, or in elderly, particularly because of abdominal discomfort and poor absorption [1–4, 49, 50]. Considering the results obtained witha low FBC-A dose, it might be beneficial in these clinical scenarios.

Our review has some limitations, mainly related to the limited number of patients enrolled in each clinical trial and the heterogeneity of these patients. Finally no studies comparing the clinical or cost effectiveness of FCB-A to FS are available.

4. Conclusions

The major challenges in the management of ID/IDA are related to the tolerability and side effects of oral iron therapy. Therefore, it is crucial to tailor the most appropriate form, dosage and duration of treatment for each patient, in order to successfully replenish iron stores.

The data presented in this short review underline the efficacy and safety of the treatment with FBC-A, and support the use of this compound in patients with ID/IDA. This review provides also preliminary evidence to suggest FBC-A as first-line treatment of ID/IDA in patients with CD. For patients with IBD further clinical trials are warranted comparing FBC-A to IV iron replacement. Finally, as it is widely assumed that poor absorption and side effects limit iron treatment compliance in several other clinical conditions as during pregnancy and in elderly, this new iron formulation seems very promisingfor the treatment of ID/IDA in these settings.

Conflict of interest

Valentina Talarico, Laura Giancotti, Mazza Giuseppe Antonio, Roberto Miniero declare no conflict of interest. Marco Bertini is R&D in the Pharmaceutical Company Laboratori Baldacci SpA.

Author details

Valentina Talarico[1*], Laura Giancotti[1,2], Giuseppe Antonio Mazza[3],
Santina Marrazzo[4], Roberto Miniero[5,6] and Marco Bertini[7]

1 Department of Pediatrics, Pugliese-Ciaccio Hospital, Catanzaro, Italy

2 Unit of Pediatrics, Magna Graecia University, Catanzaro, Italy

3 Pediatric Cardiology Unit, Città della Salute e delle Scienze, Turin, Italy

4 Unit of Ginecology, Pugliese-Ciaccio Hospital, Catanzaro, Italy

5 Department of Pediatrics, Pugliese-Ciaccio Hospital-Magna Graecia University, Catanzaro, Italy

6 Hargheisa University, Somaliland

7 R&D Department, Laboratori Baldacci SpA, Pisa, Italy

*Address all correspondence to: talaricovalentina@gmail.com

IntechOpen

References

[1] Burz C, Cismaru A, Pop V, Bojan A. Iron-deficiency anemia. In: Rodrigo L, Editor. Iron deficiency anemia. London-UK: IntechOpen; 2018; p.1-21.

[2] Camaschella C. Iron-Deficiency Anemia. Blood. 2019;133: 30-39; DOI:10.1056/NEJMra1401038.

[3] Andrews NC. Iron deficiency and related disorders. In:Greer JP, Foerster J, Rodgers GM, Paraskevas F, Glader B, Arber DA, Means RT, Editors. Wintrobe Clinical Hematology. Filadelfia: Wolters Kluwer/ Lippicott Williams&Wilkins. 2009; p 810-856.

[4] Cappellini MD, Musallam KM, Taher AT. J Iron deficiency anaemia revisited.Intern Med. 2020;287:153-170; DOI: 10.1111/joim.13004.

[5] Anrews NC, Ullrich C.K., Fleming M.D. Disorders of iron metabolism and sideroblastic anemia. In: Orkin SH, Nathan DG, Ginsburg D, Look AT, Fisher DE, Lux SE, Editors. Nathan and Oski' Hematology of infancy and childhood. Canada Sauders Elsevier. Canada. 2009; p. 521-570.

[6] Stoffel NU, Zeder C, Brittenham GM, Moretti D, Zimmermann MB. Iron absorption from supplements is greater with alternate day than with consecutive day dosing in iron-deficient anemic women. Haematologica. 2020;105:1232-1239. DOI: 10.3324/haematol. 2019.220830.

[7] Fraenkel P. Iron deficiency anemia. In:Benz EJ, Berliner NFJ. Schiffman, Editors. Anemia. Pathophysiology, Diagnosis and Management. UK. University Press 2018; p. 39-43.

[8] Gereklioglu C, Asma S, Korur A, Erdogan F, Kut A. Medication adherence to oral iron therapy in patients with iron deficiency anemia. Pak J Med Sci. 2016;32:604-607. DOI: 10.12669/ pjms.323.9799.

[9] Jeppsen RB. Toxicology and safety of Ferrochel and other iron amino acid chelates Archivos Latino americanos de Nutricion. 2001;51:126-134.

[10] Hertrampf E, Olivares M. Iron aminoacid chelates. Int J Vitam Nutr Re. 2004;74:435-443. DOI: 10.1024/ 0300-9831.74.6.435.

[11] Opinion of the Scientific Panel on Food Additives, Flavourings, Processing Aids and materials in Contact with Food on a request from the Commission related to Ferrous bisglycinate as a source of iron for use in the manufacturing of foods and in food supplements. The EFSA Journal. 2006;299:1-17.

[12] Pineda O, Ashmead HD. Effectiveness of treatment of iron deficiency anemia in infants and young children with ferrous bis-glycinate chelate. Nutrition 2001;17:381-384. DOI: 10.1016/s0899-9007(01)00519-6.

[13] Baldacci M, Gervasi GB, Bertini M. Iron deficiency anemia (ida) and iron deficiency (id): are alginates a good choise to improve oral iron bioavailability and safety? J TranslSci. 2018;4:1-3. DOI: 10.15761/JTS.1000210.

[14] Gervasi GB, Baldacci M, Bertini M. Feralgine® a new co-processed substance to improve oral iron bioavailability, taste and tolerability in iron deficiency patients. Arch Med. 2016;8:13-16.

[15] Santos D, Maurício AC, Sencadas V, Santos JD, Fernandes MH, Gomes PS. Spray Drying: An Overview. In Pignatello R, Musumeci M. Editors,

Biomaterials - Physics and Chemistry-New Edition, (December 20th 2017) IntechOpen, p. 9-35 DOI: 10.5772/intechopen.72247. Available from: https://www.intechopen.com/chapters/58222.

[16] More SK, Wagh MP. Review on spray drying technology. Int J Pharm Chem Biol Sci. 2014;4:219-225.

[17] www.fda.gov

[18] Zentilin P, Dulbecco P, Savarino E, Parodi A, Iiritano E, Bilardi C, Reglioni S., Vigneri S, Savarino V. An evaluation of the antireflux properties of sodium alginate by means of combined multichannel intraluminal impedance and pH-metry. Aliment PharmacolTher. 2005;21:29-34. DOI: 10.1111/j.1365-2036.2004.02298.

[19] Shadab MD, Singh GK, Ahuja A, Khar RK, Babota S, Sahni JK, Ali J. Mucoadhesive microspheres as a controlled drug delivery system for gastroretention. Syst Rev Pharm. 2012;3:4-14.

[20] Sachan KN, Pushkar S, Jha A, Bhattcharya A. Sodium alginate: the wonder polymer for controlled drug delivery. J Pharm Res. 2009;2:1191-1199.

[21] Ching AL, Liew CV, Chan LW, Heng PW. Modifying matrix microenvironmental pH to achieve sustained drug release from highly laminating alginate matrices. Eur J Pharm Sci. 2008;33:361-370. DOI: org/10.1016/j.jsps.2015.04.003

[22] Shadab M, Ahuja A, Khar RK, Baboota S, Chuttani K, Mishra AK, Ali J. Gastroretentive drug delivery system of acyclovir-loaded alginate mucoadhesive microspheres: formulation and evaluation. Drug Deliv. 2011;18:255-264.

[23] Gervasi GB Feralgine© powder: stereomicroscopy study. 2013 Data on file Laboratori Baldacci s.p.a.

[24] Szelaska M, Amelian A, Winnicka K. Alginate microspheres obtained by the spray drying technique as mucoadhesive carriers of ranitidine. Acta Pharm. 2015; 65:15-27. DOI: 10.1515/acph-2015-0008.

[25] Chetoni P et al. Evaluation of the dissolution process from TecnoFER Plus – July 2019 – Data on file Laboratori Baldacci SpA.

[26] Ame CA, Campa E. FERALGINE®A New Oral Iron Therapy for Iron Deficiency Anemia: Preliminary Clinical Results on a Case Series of 12 Anemic Patients Research & Reviews: Pharmacy & Pharmaceutical Sciences. 2016;5: 29-35.

[27] Mahadev S, Laszkowska M, Sundström J, Björkholm M, Lebwohl B, Green P.H.R, Ludvigsson JF. Prevalence of Celiac Disease in Patients with Iron Deficiency Anemia-A Systematic Review with Meta-analysis. Gastroenterology. 2018;155:374-382. DOI: 10.1053/j.gastro.2018.04.016.

[28] Talarico V, Giancotti L, Mazza GA, Miniero R, Bertini M. Iron Deficiency Anemia in Celiac Disease. Nutrients. 2021;13:1695-1716. DOI: 10.3390/nu13051695.

[29] Hacibekiroglu T, Akinci S, Basturk AR, Bakanay SM, Ulas T, Guney T, Dilek IA. Forgotten screening test for iron deficiency anemia: Oral iron absorbtion test. Clin Ter. 2013;164:495-97. DOI: 10.7417/CT.2013.1627.

[30] Kobune M, Miyanishi K, Takada K, Kawano Y, Nagashima H, Kikuchi S, Murase K, Iyama S, Sato T, Sato Y, Takimoto R, Kato J. Establishment of a simple test for iron absorption from the gastrointestinal tract. Int J Hematol. 2011;93:715-719. DOI: 10.1007/s12185-011-0878-8.

[31] Mazza GA, Pedrelli L, Battaglia E, Giancotti L, Miniero R. Oral iron absorption test with ferrous bisglycinate chelate in children with celiac disease: Preliminary results. Minerva Pediatr. 2019;10:139-143. DOI: 10.23736/ S0026-4946.16.04718-6.

[32] Giancotti L, Talarico V, Mazza GA, Marrazzo S, Gangemi P, Miniero R, Bertini M. Feralgine™ a New Approach for Iron Deficiency Anemia in Celiac patients. Nutrients. 2019;11:887-893. DOI: 10.3390/nu11040887.

[33] Talarico V, Giancotti L, Miniero R, Bertini M. Iron Deficiency Anemia Refractory to Conventional Therapy but Responsive to Feralgine® in a Young Woman with Celiac Disease. Int Med Case Rep J. 2021;14(2):89-93. DOI: 10.2147/IMCRJ.S291599.

[34] Rondinelli MB, Di Bartolomei A, De Rosa A, Pirelli L. Oral Iron Absorption Test (OIAT): A forgotten screening test for iron absorption from the gastrointestinal tract. A case series of Iron Deficiency Anemia (IDA) patients treated with FERALGINE®. J. Blood Disord. Med. 2017;2:1-4.

[35] Goodhand JR, Kamperidis N, Rao A, Laskaratos F, McDermott A, Wahed M, Naik S, Croft NM, Lindsay JO, Sanderson IR, Rampton DS. Prevalence and management of anemia in children, adolescents, and adults with inflammatory bowel disease. Inflamm Bowel Dis. 2012;18:513-19. DOI: 10.1002/ ibd.21740. Epub 2011 May 20.

[36] Lugg S, Beal F, Nightingale P, Bhala N, Iqbal T. Iron treatment and inflammatory bowel disease: what happens in real practice? J Crohn's Colitis. 2014;8:876-80. DOI: 10.1016/j.crohns.2014.01.011.

[37] Dignass AU, Gasche C, Bettenworth D, Birgegård G, Danese S, Gisbert JP, Gomollon F, Iqbal T, Katsanos K, Koutroubakis J, Magro F, Savoye G, Stein J, Vavricka S, on behalf of the European Crohn's and Colitis Organisation [ECCO]. European Consensus on the Diagnosis and Management of Iron Deficiency and Anaemia in Inflammatory Bowel Diseases. Journal of Crohn's and Colitis. 2015;9:211-222. DOI:10.1093/ecco-jcc/jju009.

[38] Foteinogiannopoulou K, Karmiris K, Axiaris G, Velegraki M, Gklavas A, Kapizioni K, Karageorgos C, Kateri C, Katsoula A, Kokkotis G, Koureta E, Lamouri C, Markopoulos P, Palatianou M, Pastras P, Fasoulas K, Giouleme O, Zampeli E, Theodoropoulou A, Theocharis G, Thomopoulos K, Karatzas P, Katsanos KH, Kapsoritakis A, Kourikou A, Mathou N, Manolakopoulos S, Michalopoulos G, Michopoulos S, Boubonaris A, Bamias G, Papadopoulos V, Papatheodoridis G, Papaconstantinou I, Pachiadakis I, Soufleris K, Tzouvala M, Triantos C, Tsironi E, Christodoulou DK, Koutroubakis IE, on behalf of the Hellenic group for the study of IBD The burden and management of anemia in Greek patients with inflammatory bowel disease: a retrospective, multicenter, observational study. BMC Gastroenterol. 2021; 21:269. DOI: 10.1186/s12876-021-01826-1.

[39] Vernero M, Boano V, Ribaldone DG, Pellicano R, Astegiano M Oral iron supplementation with Feralgine® in inflammatory bowel disease: a retrospective observational study. Minerva Gastroenterol Dietol. 2019;65:200-203. DOI: 10.23736/ S1121-421X.19.02572-8.

[40] Sprecacenere B, Ubezio G, Stura P, Ferrari G.M, Strada P. Uso della FERALGINA® in 52 donatori periodi di sangue anemici. Valutazione di efficacia. Blood Transuf.2018;16:s171.

[41] Butwick AJ, McDonnell N. Antepartum and postpartum anemia: a narrative review. Int J ObstetAnesth. 2021;47:102985. DOI: 10.1016/j.ijoa.2021.102985.

[42] Bertini M, Minelli F, Re P, Petruzzelli P, Menato G. Effectiveness of a new oral iron co-processed compound (FERALGINE™) in pregnant women affected by iron deficiency anemia: A prospective clinical trial. Gynecology 2019 & Dementia 2019 & Nursing Education 2019 Paris November 14-15, 2019. Androl Gynecol Curr Res. 2019;7:34.

[43] Milman N, Jønsson L, Dyre P, Pedersen PL, Larsen LG. Ferrous bisglycinate 25 mg iron is as effective as ferrous sulfate 50 mg iron in the prophylaxis of iron deficiency and anemia during pregnancy in a randomized trial. J PerinatMed. 2014;42:197-206. DOI: 10.1515/jpm-2013-0153.

[44] Russo G, Guardabasso V, Romano F, Corti P, Samperi P, Condorelli A, Sainati L, Maruzzi M, Facchini E, Fasoli S, Giona F, Caselli D, Pizzato C, Marinoni M, Boscaro G, Bertoni E, Casciana ML, Tucci F, Capolsini I, Notarangelo LD, Giordano P, Ramenghi U, Colombatti R. Monitoring oral iron therapy in children with iron deficiency anemia: an observational, prospective, multi center study of AIEOP patients. Annals of Hematology. 2020;99:413-420. DOI: 10.1007/s00277-020-03906-w.

[45] Bagna, R, Spada E, Mazzone R, Saracco P, Boetti T, Cester EA, Bertino E, Coscia A. Efficacy of Supplementation with Iron Sulfate Compared to Iron Bisglycinate Chelate in Preterm Infants. Curr Pediatr Rev. 2018;14:123-129. DOI: 10.2174/1573396314666180124101059.

[46] Liao ZC, Guan WT, Chen F, Hou DX, Wang CH, Lv YT, Qiao HZ, Chen J, Han JH. Ferrous bisglycinate increased iron transportation through DMT1 and PepT1 in pig intestinal epithelial cells compared with ferrous sulphate. J Anim Feed Sci. 2014;23:153-159. DOI: org/10.22358/jafs/65704/2014

[47] Yu X, Chen L, Ding H, Zhao Y, Feng J. Iron transport from ferrous bisglycinate and ferrous sulfate in DMT1-knockout human intestinal caco-2 cells. Nutrients. 2019;11:485. DOI:10.3390/nu11030485.

[48] Kumar A, Brookes MJ.Iron Therapy in Inflammatory Bowel Disease. Nutrients. 2020;12:3478. DOI:10.3390/nu1211347.

[49] Metzgeroth G, Hastka J. Iron deficiency anemia and anemia of chronic disorders Internist (Berl). 2015;56(9):978-8.

[50] Auerbach M, Spivak J. Treatment of Iron Deficiency in the Elderly: A New Paradigm. Clin Geriatr Med. 2019;35:307-317. DOI:10.1016/j.cger.2019.03.003.